bioanalytical chemistry

bioanalytical chemistry

Andreas Manz

Nicole Pamme

Dimitri Iossifidis

Imperial College London

Imperial College Press

Published by

Imperial College Press
57 Shelton Street
Covent Garden
London WC2H 9HE

Distributed by

World Scientific Publishing Co. Pte. Ltd.
5 Toh Tuck Link, Singapore 596224
USA office: 27 Warren Street, Suite 401-402, Hackensack, NJ 07601
UK office: 57 Shelton Street, Covent Garden, London WC2H 9HE

British Library Cataloguing-in-Publication Data
A catalogue record for this book is available from the British Library.

BIOANALYTICAL CHEMISTRY

ISBN   1-86094-370-5
ISBN   1-86094-371-3 (pbk)

Typeset by Stallion Press
Email: enquiries@stallionpress.com

Printed in Singapore by Mainland Press

CONTENTS

Preface   ix
List of Abbreviations   xi

Chapter 1   Biomolecules   1

1.1   Amino Acids, Peptides and Proteins   1
        1.1.1   Amino Acids   2
        1.1.2   Peptides and Proteins   7
1.2   Nucleic Acids   14
        1.2.1   The Structure of Nucleic Acids   15
        1.2.2   Synthesis of Proteins   20
1.3   Biomolecules in Analytical Chemistry   22
        1.3.1   Classical Analytical Chemistry   22
        1.3.2   Limitations of Classical Analytical Chemistry   22
        1.3.3   Bioanalytical Chemistry   23

Chapter 2   Chromatography   29

2.1   The Principle of Chromatography   29
2.2   Basic Chromatographic Theory   31
2.3   Application of Liquid Chromatography for Bioanalysis   34
        2.3.1   Reversed Phase Liquid Chromatography (RP-LC)   34
        2.3.2   Ion Exchange Chromatography (IEC)   37
        2.3.3   Affinity Chromatography   40
        2.3.4   Size Exclusion Chromatography (SEC)   42

Chapter 3   Electrophoresis   47

3.1   Principle and Theory of Electrophoresis   48
        3.1.1   Electrophoretic Mobility   49
        3.1.2   Joule Heating   50
        3.1.3   Electroosmotic Flow (EOF)   50
        3.1.4   Separation Efficiency and Resolution   54
3.2   Gel Electrophoresis (GE)   56
        3.2.1   Instrumentation for Gel Electrophoresis   57
        3.2.2   Modes of Gel Electrophoresis   63
        3.2.3   Sodium Dodecyl Sulphate–Polyacrylamide Gel
                Electrophoresis (SDS–PAGE)   63

		3.2.4	Isoelectric Focussing (IEF)	64
		3.2.5	Two-Dimensional Gel Electrophoresis (2D-GE)	67
3.3	Capillary Electrophoresis (CE)			69
		3.3.1	Capillary Electrophoresis Instrumentation	70
		3.3.2	Capillary Zone Electrophoresis (CZE)	75
		3.3.3	Capillary Isoelectric Focussing (CIEF)	76
		3.3.4	Micellar Electrokinetic Chromatography (MEKC)	77
		3.3.5	Capillary Gel Electrophoresis (CGE)	82

Chapter 4   Mass Spectrometry   **85**

4.1	The Principle of Mass Spectrometry			85
		4.1.1	Ionisation	86
		4.1.2	Mass Analyser	86
		4.1.3	Detector	87
4.2	Matrix Assisted Laser Desorption Ionisation – Time of Flight Mass Spectrometry (MALDI-TOF/MS)			87
		4.2.1	Ionisation Principle	87
		4.2.2	Mass Analysis in Time-of-Flight Analyser	90
		4.2.3	Detection of Ions	92
		4.2.4	Resolution	92
		4.2.5	Sample Pretreatment	93
		4.2.6	Applications of MALDI	94
4.3	Electrospray Ionisation Mass Spectrometry (ESI-MS)			97
		4.3.1	Ionisation Principle	98
		4.3.2	ESI – Source and Interface	99
		4.3.3	Quadrupole Analyser	100
		4.3.4	Applications of ESI-MS	101

Chapter 5   Molecular Recognition:
Bioassays, Biosensors, DNA-Arrays and
Pyrosequencing   **109**

5.1	Bioassays			110
		5.1.1	Antibodies	111
		5.1.2	Antigens	113
		5.1.3	Antibody-Antigen Complex Formation	114
		5.1.4	Assay Formats	115
		5.1.5	Home Pregnancy Test	120
		5.1.6	Enzyme Immunoassays (EI and ELISA)	121
5.2	Biosensors			125
		5.2.1	Bioreceptors	126

	5.2.2	Transducers	127
	5.2.3	The Blood Glucose Sensor	128
5.3	DNA Binding Arrays		131
	5.3.1	The Principle of DNA Arrays	131
	5.3.2	Fabrication of DNA Arrays	132
	5.3.3	Development and Analysis of a DNA Array	134
	5.3.4	DNA Sequencing with Arrays	134
	5.3.5	Other Applications of DNA Arrays	136
5.4	DNA Identification by Pyrosequencing		136
	5.4.1	The Principle of Pyrosequencing	137
	5.4.2	Sample Preparation and Instrumentation	140
	5.4.3	Applications of Pyrosequencing	140

Chapter 6 Nucleic Acids: Amplification and Sequencing

Chapter 6 Nucleic Acids: Amplification and Sequencing 143

6.1	Extraction and Isolation of Nucleic Acids		143
	6.1.1	CsCl Density Gradient Centrifugation	144
	6.1.2	Total Cellular DNA Isolation	145
	6.1.3	RNA Isolation – The Proteinase K method	145
6.2	Nucleic Acid Amplification – The Polymerase Chain Reaction (PCR)		146
	6.2.1	The Principle of PCR	146
	6.2.2	The Rate of Amplification During a PCR	149
	6.2.3	Reagents for PCR	151
	6.2.4	Real-Time PCR	153
	6.2.5	Reverse Transcription – PCR (RT-PCR)	155
6.3	Nucleic Acid Sequencing		156
	6.3.1	The Use of Restriction Enzymes in Sequencing	156
	6.3.2	The Chemical Cleavage method (The Maxam-Gilbert method)	158
	6.3.3	The Chain Terminator method (The Sanger or Dideoxy method)	162
6.4	RNA Sequencing		166

Chapter 7 Protein Sequencing

Chapter 7 Protein Sequencing 169

7.1	Protein Sequencing Strategy		170
7.2	End-group Analysis		170
	7.2.1	*N*-terminal Analysis (Edman Degradation)	171
	7.2.2	*C*-terminal Analysis	172
7.3	Disulfide Bond Cleavage		175

7.4   Separation and Molecular Weight Determination of the Protein
       Subunits   177
7.5   Amino Acid Composition   178
7.6   Cleavage of Specific Peptide Bonds   179
       7.6.1   Enzymatic Fragmentation   180
       7.6.2   Chemical Fragmentation Methods   183
7.7   Sequence Determination   183
7.8   Ordering of Peptide Fragments   186
7.9   Determination of Disulfide Bond Positions   186
7.10   Protein Sequencing by Mass Spectrometry   187

Index   **189**

Preface

In a time when sequencing the human genome has just recently been completed, when Nobel prizes are awarded to inventors of bioanalytical instrumentation and when the reading of journals such as *Science* or *Nature* has become ever more difficult to the chemist due to the flood of molecular biology terminology appearing in these groundbreaking publications ... At exactly this time, it seems imperative to provide a small introductory textbook covering the most frequently used instrumental methods of analytical chemistry in molecular biology. The increasingly interdisciplinary nature of modern research makes it essential for researchers of different backgrounds to have at least a minimal understanding of neighbouring sciences if they are to communicate effectively.

For many years, Professor Manz has presented a "bioanalytical chemistry" course at Imperial College, whilst being acutely aware of the lack of a suitable textbook for this subject. Of course, each individual subunit could be found in yet another biochemistry, mass spectrometry, separations or analytical chemistry textbook. However, considering the importance of biomolecules in recent academic and industrial research, it is somewhat surprising that this is not yet reflected in current analytical chemistry textbooks. In the light of these facts, it seems appropriate for us to write a new book concerning the various aspects of biomolecular analysis.

This book is aimed primarily at chemistry students, but is also intended to be a useful reference for students, lecturers and industrial researchers in biological and medicinal sciences who are interested in bioanalysis techniques. It is assumed that the basic principles and instrumental techniques of analytical chemistry are already common knowledge. An important objective of this book is to give an appreciation of how analytical methods are influenced by the properties that are peculiar to biomolecules. The priorities that govern the choice of instrumental techniques for the analysis of molecules such as DNA and proteins are radically different to those applicable to classical analytical chemistry (see Summary of Chapter 1). Whereas samples containing small molecules can be characterised by gas or liquid chromatography, when it comes to DNA sequencing or proteomic analysis, there is a sudden need for sheer separation power. Hence, students must have as clear an understanding of isoelectric focussing or 2D slab gel separation as they would of conventional chromatography. Other methods described in this book may be completely new to the chemist. For example, the polymerase chain reaction used for DNA amplification or the Sanger reaction for DNA sequencing, where low yield chemical reactions are performed to generate hundreds of products.

In the first chapter of this book, a general introduction to biomolecules is given. This is followed by several chapters describing various instrumental techniques and bioanalytical methods. These include: electrophoresis, isoelectric focussing, MALDI-TOF, ESI-MS, immunoassays, biosensors, DNA arrays, PCR, DNA and protein sequencing. Instead of being a comprehensive reference or textbook, it is intended that this book should provide introductory reading, perhaps alongside a taught course. A list of references is given at the end of each chapter, should further information be required on any particular subject.

Hopefully, this book will be well received by both teachers and students, particularly in a time when techniques of bioanalysis should be familiar to every chemistry graduate.

The authors would like to thank Dr. Alexander Iles for his comments on the manuscript.

Andreas Manz, Nicole Pamme, Dimitri Iossifidis
London, March 12, 2003

List of Abbreviations

2D-GE	two-dimensional gel electrophoresis
A	Adenine
α	selectivity factor
Ab	antibody
ABTS	2,2′-azino-bis (ethyl-benzothiazoline-6-sulfonate)
ac	alternating current
α-CHCA	α-cyano-4-hydroxy-cinnamic acid
AChE	acetylcholine esterase
ADT	adenosine diphosphate
Ag	antigen
AIDS	acquired immunodeficiency syndrome
Ala	Alanine
AMP	adenosine monophosphate
AN	aggregation number
AP	alkaline phosphatase
APS	adenosine phosphosulphate
Arg	Arginine
Asn	Asparagine
Asp	Aspartic acid
ATP	adenosine triphosphate
bp	base pair
BSA	bovine serum albumin
c	concentration
C	Cytosine
C%	degree of cross-linking
CCD	charged coupled device
cDNA	complementary DNA
CE	capillary electrophoresis
CGE	capillary gel electrophoresis
CHAPS	3-[(cholamido propyl) dimethyl ammonio]-1-propane sulphonate
CI	chemical ionisation
CID	collision-induced dissociation
CIEF	capillary isoelectric focussing
CM	carboxy methyl
CMC	critical micelle concentration

CNBr	cyanogen bromide
CTAB	cetyltrimethylammonium bromide
CTAC	cetyltrimethylammonium chloride
Cys	Cysteine
CZE	capillary zone electrophoresis
D	diffusion coefficient
Da	Dalton
DAD	diode array detector
dATP	deoxyadenine triphosphate
dATP-αS	deoxyadenine α-thio-triphosphate
dc	direct current
dCTP	deoxycytosine triphosphate
ddNTP	2',3'-dideoxynucleotide triphosphate
DEAE	diethyl aminoethyl
dGTP	deoxyguanine triphosphate
DHBA	2,5-dihydroxy benzoic acid
DMS	dimethyl sulphate
DMSO	dimethyl sulphoxide
DNA	deoxyribonucleic acid
dNDP	deoxynucleotide diphosphate
dNMP	deoxynucleotide monophosphate
dNTP	deoxynucleotide triphosphate
DoTAB	dodecyl trimethyl ammonium bromide
ΔpI	resolution (in isoelectric focusing)
dsDNA	double stranded DNA
DTT	dithiothreitol
dTTP	deoxythymine triphosphate
ϵ	dielectric constant
E	electric field strength
e	electron charge
EI	electron impact ionisation
EI	enzyme imunoassay
E_{kin}	kinetic energy
ELISA	enzyme-linked immunosorbent assay
EOF	electroosmotic flow
ESI	electrospray ionisation
Fab	antigen binding fragment of Ig
FAB	fast atom bombardment
Fc	crystallisable fragment of Ig
F_{ef}	electric force
F_{fr}	frictional force
FRET	fluorescence resonance energy transfer
FWHM	full width at half maximum

G	Guanine
GC	gas chromatography
GE	gel electrophoresis
Gln	Glutamine
Glu	Glutamic acid
Gly	Glycine
GOx	glucose oxidase
GPC	gel permeation chromatography
H	height equivalent of a theoretical plate
η	viscosity
hCG	human chorionic gonadotropin
His	Histidine
HIV	human immunodeficiency virus
HPCE	high performance capillary electrophoresis
HPG	human genome project
HPLC	high performance liquid chromatography
HRP	horseradish peroxidase
i.d.	inner diameter
IEC	ion exchange chromatography
IEF	isoelectric focussing
Ig	Immunoglobulin
Ile	Isoleucine
IPG	immobilised pH gradient
IR	infrared
k'	capacity factor
k_3	turnover of an enzyme
K_{eq}	equilibrium constant of antibody-antigen complex formation
K_m	Michaelis-Menten constant
L	length (of capillary, colum or gel)
λ	wavelength
LC	liquid chromatography
Leu	Leucine
LIF	laser induced fluorescence
Lys	Lysine
m	mass
M	molar, $mol\ L^{-1}$
m/z	mass-to-charge ratio
MALDI	matrix assisted laser desorption ionisation
μ_{app}	apparent mobility
MECC	micellar electrokinetic capillary chromatography
MEKC	micellar electrokinetic chromatography
μ_{EOF}	electroosmotic mobility
μ_{ep}	electrophoretic mobility

$\mu_{ep,AVE}$	average electrophoretic mobility of two analytes
Met	Methionine
mM	millimolar
mRNA	messenger RNA
MS	mass spetrometry
MS/MS	tandem mass spectrometry
μ_{tot}	total mobility
MW	molecular weight
N	plate number
N_0	initial number of DNA molecules in PCR
N_m	number of DNA molecules in PCR
NMR	nuclear magnetic resonance
ODS	octadecyl silane
OPA	ortho-phthalaldehyde
ox.	oxidised
PA	polyacrylamide
PAGE	polyacrylamide gel electrophoresis
PCR	polymerase chain reaction
PEG	polyethylene glycol
pH	potentium hydrogenis
Phe	Phenylalanine
pI	isoelectric point
PICT	phenylisothiocyanate
pK	dissociation constant
ppb	parts per billion
PPi	pyrophosphate
ppm	parts per million
RP	reversed phase
Pro	Proline
PSD	post source decay
PTH	phenylthiohydantoin
q	charge of ion
QT-PCR	quantitative PCR
r	ionic/molecular radius
red.	reduced
RNA	ribonucleic acid
R_S	resolution
RT	reverse transcription
RT-PCR	reverse transcription polymerase chain reaction
s	signal intensity
s^2	peak dispersion
SA	sinapinic acid

SC	sodium cholate
SDS	sodium dodecyl sulphate
SEC	size exclusion chromatography
Ser	Serine
SLD	soft laser desorption
SNP	single nucleotide polymorphism
ssDNA	single stranded DNA
STC	sodium taurocholate
STS	sodium tetradecyl sulphate
t	migration time
T	Thymine
T%	total gel concentration
t_0	zero retention time
Taq	Thermus aquaticus
TFA	trifluoroacetic acid
Thr	Threonine
t_{mc}	retention time of micelles
TOF	time of flight
t_R	retention time
TRIS	tris (hydroxylmethyl)-aminomethane
tRNA	transfer RNA
Trp	Tryptophan
Tyr	Tyrosine
u	flow rate
U	Uracil
UV	ultraviolet
V	applied voltage
v	migration velocity
V_0	inter particle volume
Val	Valine
v_{EOF}	velocity of electroosmotic flow
v_{ep}	electrophoretic velocity
V_g	volume of gel particles
V_i	intrinsic volume
vis	visible
v_{MC}	velocity of micelles
V_R	retention volume
V_t	total volume
w	peak width
z	ion charge
ζ	zeta potential

Chapter 1

BIOMOLECULES

In this chapter, you will learn about...

♦ ... the biomolecules that are most commonly analysed in bioanalytical chemistry: amino acids, proteins and nucleic acids.

♦ ... the structure of these biomolecules and their physical and chemical characteristics.

♦ ... some of the functions of these biomolecules and how they interact with each other in the cell.

Chemists are likely to be familiar with certain biomolecules such as carbohydrates and lipids from their organic chemistry lectures. However, many do not have a clear understanding of the composition and function of other biomolecules such as proteins and DNA. This chapter introduces the biomolecules, which are the target of the analytical methods described in the following chapters.

1.1 Amino Acids, Peptides and Proteins

Amino acids are the building blocks for peptides and proteins and play an important part in metabolism. 20 different amino acids are found in living organisms. They can connect to each other via peptide bonds to form long chains. Proteins may consist of thousands of amino acids and can have molecular weights of up to several million Dalton (Da). Shorter chains of up to a few hundred amino acids are referred to as peptides. The sequence of the amino acids within the molecule is essential for the structure and function of proteins and peptides in biological processes.

1.1.1   *Amino Acids*

The general structure of an amino acid is shown in Fig. 1.1. It consists of a tetrahedral carbon atom (C-alpha) connected to four groups: a basic amino group ($-NH_2$),
an acidic carboxyl group ($-COOH$), a hydrogen atom ($-H$) and a substituent group
($-R$), which varies from one amino acid to another. The amino group is in the alpha
position relative to the carboxyl group, hence the name *α-amino acids*. Amino acids
are chiral with the exception of glycine, where the R substituent is a hydrogen atom.
All natural amino acids have the same absolute configuration: the L-form in the
Fischer convention or the S-form according to the Cahn-Ingold-Prelog rules, with
the exception of cysteine, which has the R-configuration.

Amino acids can be classified according to their substituent R groups (Fig. 1.2 to
Fig. 1.8): in *basic amino acids*, R contains a further amino group, whereas in *acidic
amino acids,* R contains a further carboxyl group. In addition, there are *aliphatic*,
aromatic, *hydroxyl containing* and *sulfur containing amino acids* according to the
nature of the substituent, as well as a *secondary* amino acid.

For convenience, the names for amino acids are often abbreviated to either a
three symbol or a *one symbol short form*. For example, Arginine can be referred

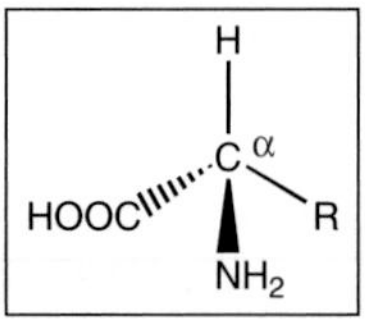

Fig. 1.1.   General structure of an α-L-amino acid.

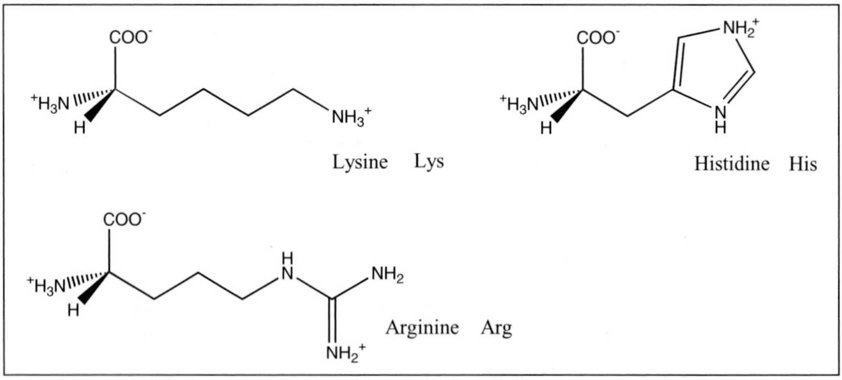

Lysine   Lys

Histidine   His

Arginine   Arg

Fig. 1.2. Basic amino acids.

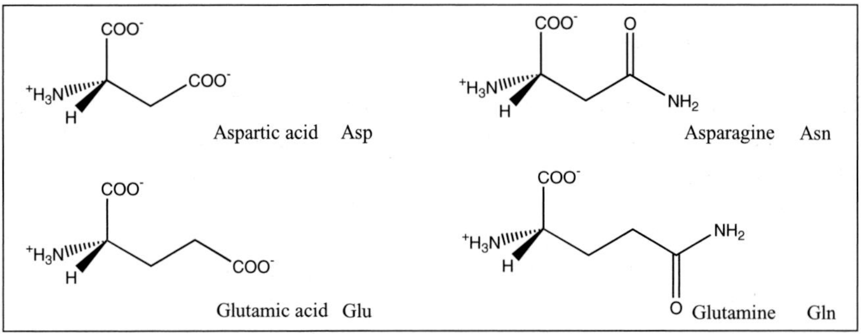

Fig. 1.3. Acidic amino acids.

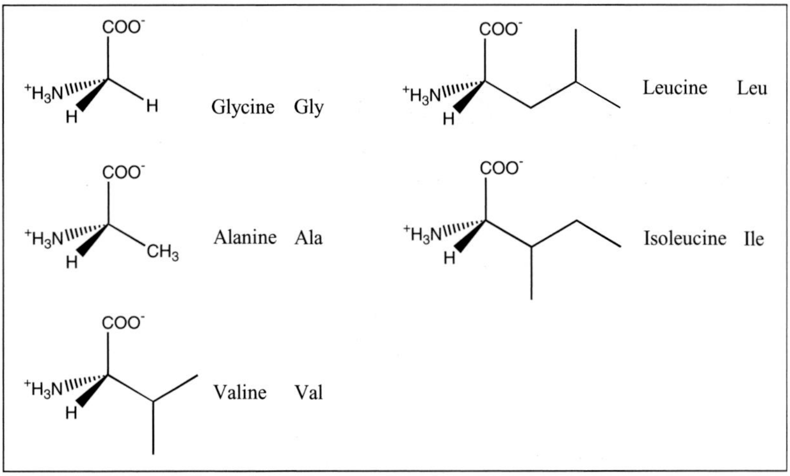

Fig. 1.4. Aliphatic amino acids.

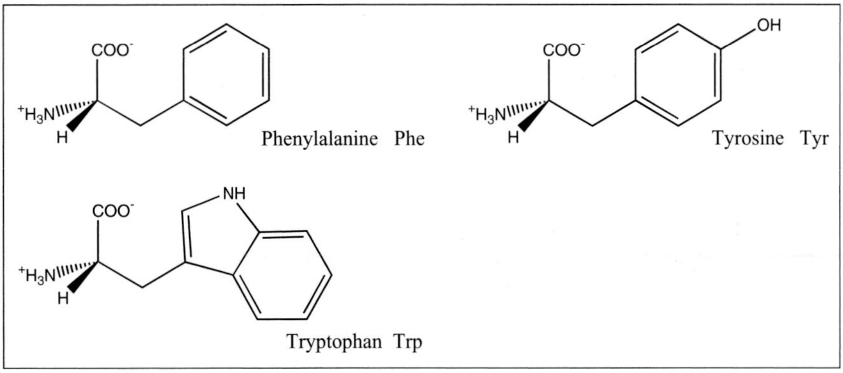

Fig. 1.5. Aromatic amino acids.

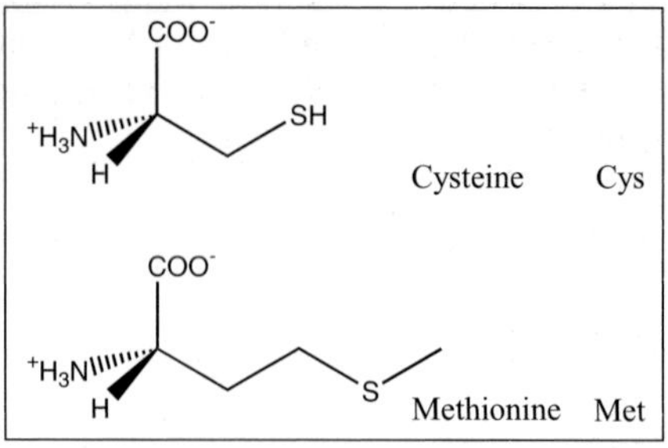

Fig. 1.6. Sulfur containing amino acids.

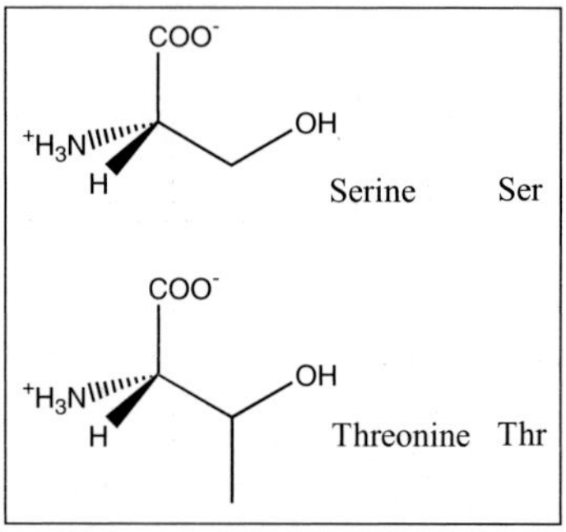

Fig. 1.7. Amino acids with an alcoholic hydroxyl group.

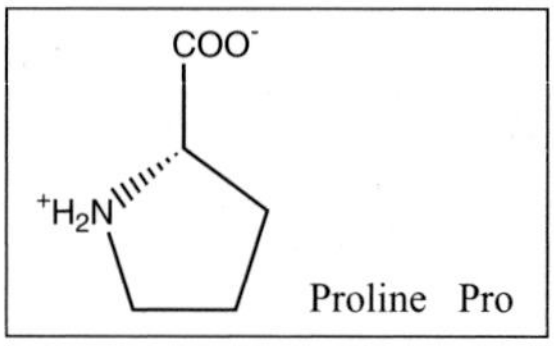

Fig. 1.8. Secondary amino acid.

to as Arg or R and Glycine can be shortened to Gly or G. The abbreviations for the 20 natural amino acids are listed in Table 1.1. These naturally occurring amino acids are the building blocks of peptides and proteins. Any particular amino acid is not likely to exceed 10 % of the total composition of a protein (see Table 1.1).

Amino acids can also be classified according to their polarity and charge at pH 6 to 7, which corresponds to the pH range found in most biological systems. This is often referred to as the *physiological pH*. *Non-polar amino acids* with no

Table 1.1. Natural amino acids.

Name	Three and one letter symbols		M_r (Da)	found[1] (%)	$pK_1^{(2)}$ α-COOH	$pK_2^{(2)}$ α-NH$_3^+$	$pK_R^{(2)}$ side-chain
basic amino acids							
Lysine	Lys	K	146.2	5.9	2.16	9.06	10.54 ε-NH$_3^+$
Histidine	His	H	155.2	2.3	1.8	9.33	6.04 imidazole
Arginine	Arg	R	174.2	5.1	1.82	8.99	12.48 guanidino
acidic amino acids							
Aspartic acid	Asp	D	133.1	5.3	1.99	9.90	3.90 β-COOH
Glutamic acid	Glu	E	147.1	6.3	2.10	9.47	4.07 γ-COOH
Asparagine	Asn	N	132.1	4.3	2.14	8.72	
Glutamine	Gln	Q	146.2	4.3	2.17	9.13	
aliphatic amino acids							
Glycine	Gly	G	75.1	7.2	2.35	9.78	
Alanine	Ala	A	89.1	7.8	2.35	9.87	
Valine	Val	V	117.2	6.6	2.29	9.74	
Leucine	Leu	L	131.2	9.1	2.33	9.74	
Isoleucine	Ile	I	131.2	5.3	2.32	9.76	
aromatic amino acids							
Phenylalanine	Phe	F	165.2	3.9	2.20	9.31	
Tyrosine	Tyr	Y	181.2	3.2	2.20	9.21	10.46 phenol
Trytophan	Trp	W	204.2	1.4	2.46	9.41	
sulfur containing amino acids							
Cysteine	Cys	C	121.2	1.9	1.92	10.70	8.37 sulfhydryl
Methionine	Mel	M	149.2	2.2	2.31	9.28	
amino acids with alcoholic hydroxyl groups							
Serine	Ser	S	105.1	6.8	2.19	9.21	
Threonine	Thr	T	119.1	5.9	2.09	9.10	
amino acid with secondary amino group							
Proline	Pro	P	115.1	5.2	1.95	10.64	

Sources:
(1) R. F. Doolittle, *Database of nonredundant proteins,* in G. D. Fasman (Ed.), *Predictions of Protein Structure and the Principles of Protein Conformation*, Plenum Press, 1989.
(2) R. M. C. Dawson, D. C. Elliott, W. H. Elliott, K. M. Jones, *Data for Biochemical Research,* 3rd edition, Oxford Science Publications, 1986.

net charge are Alanine, Valine, Leucine, Isoleucine, Phenylalanine, Tryptophan, Methionine and Proline. *Polar amino acids* have no net charge but carry a polar group in the substituent R. Glycine, Asparagine, Glutamine, Tyrosine, Cysteine, Serine and Threonine fall into this category. *Positively charged amino acids* at physiological pH are Lysine, Histidine and Arginine; whereas *negatively charged amino acids* are Aspartic acid and Glutamic acid.

In addition to the 20 natural amino acids, there are other amino acids, which occur in biologically active peptides and as constituents of proteins. These will not be covered in this textbook.

1.1.1.1 *Zwitterionic character, pK and pI*

As amino acids contain a basic and an acidic functional group, they are *amphoteric*. The carboxyl group of an amino acid has a pK between 1.8 and 2.5, the amino group has a pK between 8.7 and 10.7 (see Table 1.1). At the pH found under physiological conditions, pH 6 to 7, the amino group is ionised to $-NH_3^+$ and the carboxyl group is ionised to $-COO^-$. Hence, at physiological pH amino acids are *zwitterionic*. At low pH values, the carboxyl group is protonated to $-COOH$ and the amino acid becomes positively charged. At high pH values, the amino group is deprotonated to $-NH_2$ and the amino acid becomes negatively charged (Fig. 1.9). Functional groups in the substituents may have different pK values as well (see Table 1.1).

For every amino acid, there is a specific pH value at which it exhibits no net charge. This is called the *isoelectric point, pI*. At its isoelectric point, an amino acid remains stationary in an applied electric field, i.e. it does not move to the positive or negative pole. The isoelectric point can be estimated via the *Henderson-Hasselbalch equation*:

$$pI = \frac{1}{2}(pK_i + pK_j) \qquad \text{(equation 1.1)}$$

where pK_i and pK_j are the dissociation constants of the ionisation steps involved. This calculation is straightforward for mono-amino and mono-carboxylic acids, where pK_i and pK_j are the pK values of the amino group and the carboxylic group, respectively. For amino acids with ionisable side chains, the calculation of the pI value is more complex. The pI values for the natural amino acids are listed in Table 1.2, and in Table 1.3 pI values are given for some proteins. Differences in pI can be utilised to separate amino acids or proteins in an electric field. This technique is called isoelectric focussing and will be discussed in detail in sections 3.2.4 and 3.3.3.

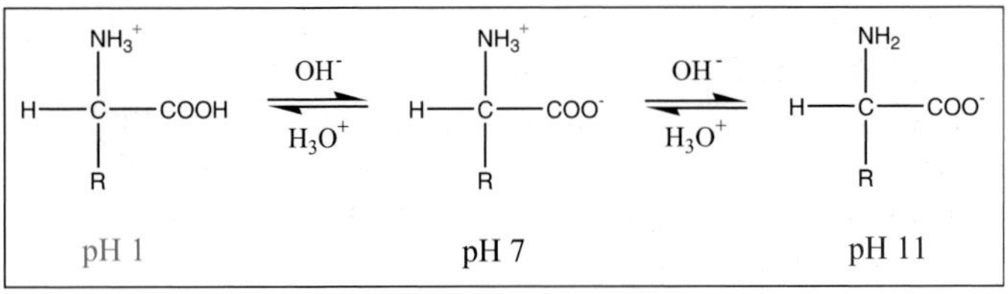

Fig. 1.9. Charge of an amino acid at different pH values: zwitterionic character at pH 7, positive charge at low pH and negative charge at high pH.

Table 1.2. pI values of natural amino acids.

Amino acids Non-polar chain	pI	Amino acids Polar chain	pI	Amino acids Charged chain	pI
Alanine	6.02	Glycine	5.97	Lysine	9.74
Valine	5.97	Asparagine	5.41	Histidine	7.58
Leucine	5.98	Glutamine	5.65	Arginine	10.76
Isoleucine	6.02	Tyrosine	5.65	Aspartic acid	2.87
Phenylalanine	5.98	Cysteine	5.02	Glutamic acid	3.22
Tryptophan	5.88	Serine	5.68		
Methionine	5.75	Threonine	6.53		
Proline	6.10				

Table 1.3. pI values of some proteins.

Protein	pI	Protein	pI
Pepsin	<1.0	Myoglobin (horse)	7.0
Ovalbumin (hen)	4.6	Haemoglobin (human)	7.1
Serum albumin (human)	4.0	Ribonuclease A (bovine)	7.8
Tropomyosin	5.1	Cytochrome c (horse)	10.6
Insulin (bovine)	5.4	Histone (bovine)	10.8
Fibrinogen (human)	5.8	Lysozyme (hen)	11.0
γ-Globuline (human)	6.6	Salmine (salmon)	12.1
Collagen	6.6		

1.1.2 *Peptides and Proteins*

Peptides and proteins are *macromolecules* made up from long chains of amino acids joined head-to-tail via *peptide bonds*. The three-dimensional structure of a protein is very well defined and is essential for it to function. Proteins are found

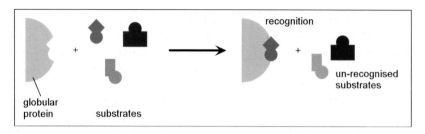

Fig. 1.10. Globular proteins like enzymes and antibodies have a specific surface that recognises only specific substrates.

in all forms of living organisms and perform a wide variety of tasks. The function and structure of proteins are outlined in the following sections.

1.1.2.1   *The biological function of proteins*

In general, there are two types of protein structures: (1) *fibrous*, elongated proteins which are not soluble in water and provide structural support and (2) *globular* spherical proteins which are water soluble and have specific functions in the immune system and metabolism.

   Globular proteins have a compact, spherical structure with very characteristic grooves and peaks on their surface. Analogous to a key fitting into a lock, other molecules fit into these grooves and peaks. This makes globular proteins *specific* when it comes to interacting with or recognising other molecules (Fig. 1.10). *Enzymes* are an example of such specific proteins. They are biochemical catalysts, which lower the activation energy and, thus, accelerate immensely the reaction rate of biological reactions. An enzyme can only react with a substrate if the location of its functional groups and hydrogen bonds as well as its shape matches the active site of the enzyme. Ribonuclease for example is an enzyme secreted by the pancreas to specifically digest ribonucleic acid (RNA). *Antibodies* are another example of highly specific globular proteins. They can recognise intruders, *antigens*, and bind to them in a *key-lock mechanism*. Enzymes and antibodies are used as molecular recognition elements in bioassays (section 5.1) and biosensors (section 5.2).

   In the body, proteins also function as *transport* and *storage media*. For example, haemoglobin is responsible for the transport of oxygen in the blood stream, transferrin for the transport of iron. Ferritin is an example of a protein with a storage function, which can be found in the liver. It forms a complex with iron, and thus binds and stores the metal. In the form of *hormones*, polypeptides can also act as *chemical messengers*. By interacting with a matching receptor, usually found in the cell membrane, they regulate a wide variety of tasks in metabolism. For example, three hormones found in the pancreas, glucagon, insulin and somato-statin, regulate the storage and release of glucose and fatty acids. Other hormones control digestion, growth and cell differentiation. Hormones form a large class of chemical substances. Most hormones are polypeptides, however, some are amino acid derivates or steroids.

Fibrous proteins have a high tensile strength and mechanical stability. Their function is to provide *structural support* to tissues. Collagen, for example, gives connective strength to skin, bones, teeth and tendons. Ceratin is the major component of hair and nails.

1.1.2.2   *The structure of proteins*

Proteins are not just randomly coiled chains of amino acids. A variety of intramolecular interactions enables the amino acid chain to fold in a specific way to give the protein a three-dimensional structure and shape. This structure is crit-ical for its activity and function. Several amino acid strings can be entangled and connected to each other via *disulfide bridges*. Parts of the amino acid chain can be organised into helices or sheets. Globular proteins like enzymes and antibodies are more folded and coiled whereas fibrous proteins are more filamentous and elon-gated. To describe the complex structure of proteins, four levels of organisation are distinguished: *primary, secondary, tertiary* and *quaternary* structures.

Primary structure

The sequence of amino acids determines the primary structure of a protein. Chang-ing just a single amino acid in a critical position of the protein can significantly alter its activity and function and be the cause of disease and disorders. The amino acids are connected to each other in a head-to-tail fashion by formation of a *peptide bond* (Fig. 1.11), the condensation of a carboxylic and an amino group with the elimination of water.

Two amino acids connected via a peptide bond are called a *dipeptide*, three acids a *tripeptide* and so on. With an increasing number of acids in the sequence, the molecules are referred to as *oligopeptides* and *polypeptides*. The C−N bond cannot

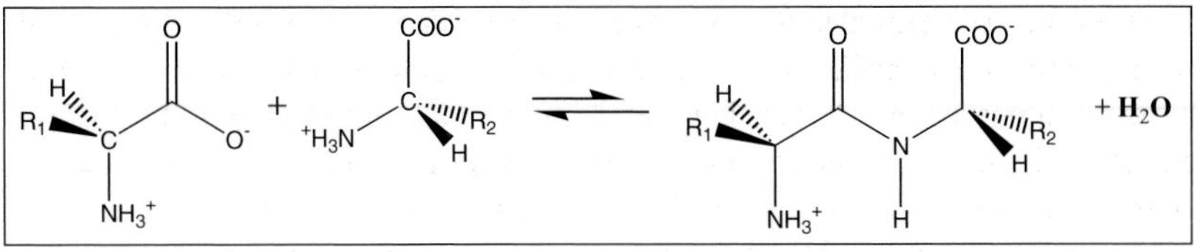

Fig. 1.11. Peptide bond formation from two amino acids.

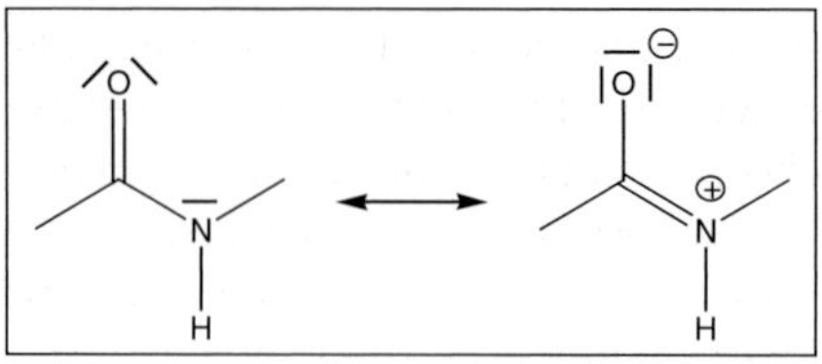

Fig. 1.12. Double bond character of the C–N bond in a peptide.

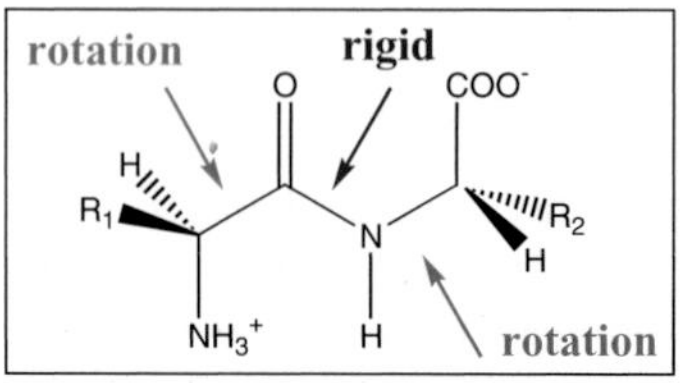

Fig. 1.13. The C–N bond is rigid due to the partial double bond character, rotation is possible within steric constraints around the bonds to the α C-atoms.

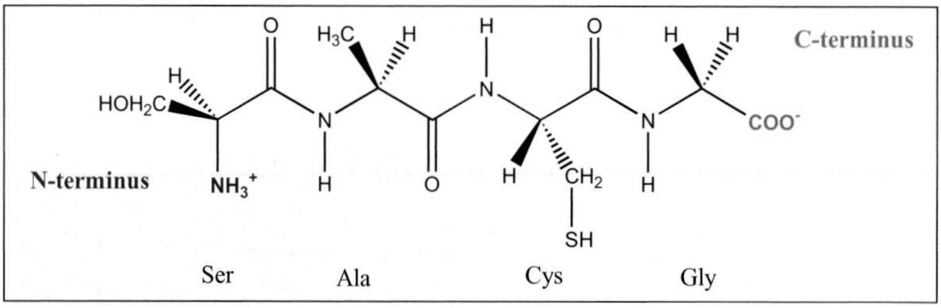

Fig. 1.14. A peptide with the amino acid sequence Ser-Ala-Cys-Gly showing N-terminus and C-terminus.

rotate due to its partial double bond character (Fig. 1.12). Hence, the peptide unit NH–CO is rigid. The bonds to the neighbouring alpha C-atoms can rotate within steric constraints (Fig. 1.13) and play an important part in folding of the protein. The peptide units together with the tetrahedral C-atoms form the *backbone* of a protein, while the R substituents are referred to as *side chains*.

An example of a peptide consisting of four amino acid residues (Ser-Ala-Cys-Gly) is given in Fig. 1.14. To be unambiguous about start and end of a sequence, the first amino acid residue is always the one with the free amino group, the *N-terminus*, and is written to the left. The last amino acid in the chain is the *C-terminus* with the free carboxyl group and is written to the right.

Peptides can also have a circular structure, i.e. they "bite their own tail". An example of such a peptide is the potassium carrier, valinomycin.

With the 20 naturally occurring L-amino acids, it is possible to form an immense number of combinations and permutations. For a dipeptide there are already $20^2 = 400$ possible arrangements, for a tripeptide $20^3 = 8,000$. A relatively small protein with 100 amino acid residues can be arranged in $20^{100} = 1.27 \times 10^{130}$ different ways, an enormous number, especially when bearing in mind that there are "only" 10^{78} atoms in the whole universe. The bioanalytical chemist has to face a difficult task, if he wants to determine the exact sequence of amino acids in a protein. Nonetheless, their analysis has become commonplace and the methods involved are discussed in chapter 7.

Secondary structure

Secondary structures are regular elements such as *α-helices* and *β-pleated sheets*, which are formed between relatively small parts of the protein sequence. These structural domains are determined by the conformation of the peptide backbone, the influence of side-chains is not taken into account for secondary structures.

An *α-helix* (Fig. 1.15) is a right-handed coil, which is held together by hydrogen bonding between a –CO group of an nth amino acid residue in the sequence and the –NH group of the $n+4^{th}$ amino acid residue. The coiling is such that the –R groups

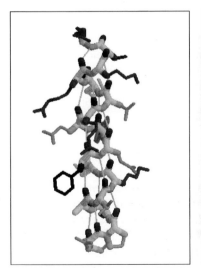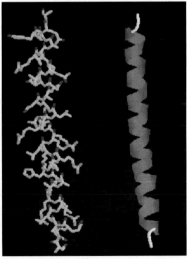

Fig. 1.15. Left: Structure of an α-helix with the –R substituents pointing outwards. Right: Schematic drawing of an α-helix as commonly used in drawings of proteins.

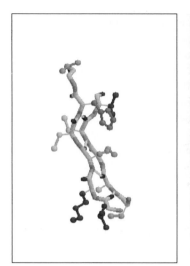 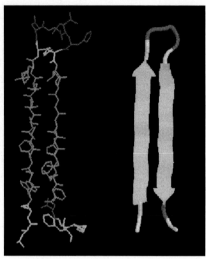

Fig. 1.16. Left: Structure of a β-pleated sheet with the –R substituents pointing outwards. Right: Schematic drawing of a β-pleated sheet as commonly used in drawings of proteins.

are pointing outwards perpendicular to the axis of the coil. α-helices are important in structural proteins like ceratin. Not all amino acids favour α-helix formation due to steric hindrance. The secondary amino acid proline, for example, is likely to disrupt the formation of a helix.

In a *β-pleated sheet* (Fig. 1.16), two polypeptide backbones are folded and aligned next to each other. They are connected via hydrogen bonds. The amino acid substituents R are pointing outwards to the top or bottom of the sheet. Adjacent chains can be aligned either in the same direction (parallel β-folding) or in opposite directions (antiparallel β-folding), as shown in Fig. 1.16. β-folding often occurs with amino acids carrying small non-charged side chains.

Tertiary structure

The tertiary structure describes the complete three-dimensional structure of the whole polypeptide chain. It includes the relationship of different domains formed by the protein's secondary structure and the interactions of the amino acid substituent –R groups. An example of a protein chain with α-helices and β-folding, the enzyme ribonuclease, is shown in Fig. 1.17. The specific folding of a protein is only thermodynamically stable within a restricted range of environmental parameters, i.e. the right temperature, pH and ionic strength. Outside of this range, the protein could unfold and lose its activity.

Fig. 1.17. 3D-structure of ribonuclease H from *Escherichia coli* with α-helices and β-folding.

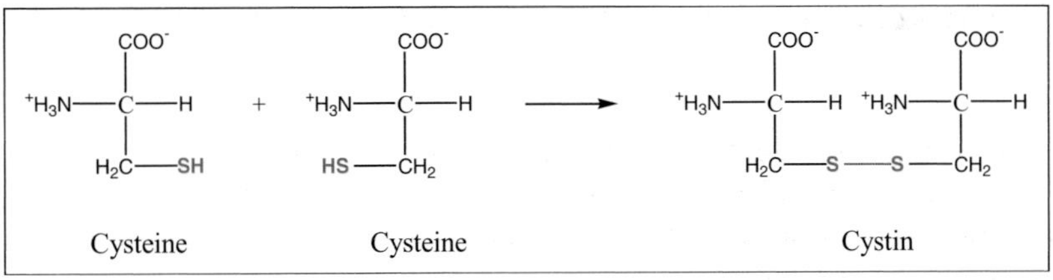

Fig. 1.18. Formation of a disulfide bridge.

Quaternary structure

A protein can consist of two or more separate polypeptide chains linked together. Other, non-amino acid components such as minerals, lipids and carbohydrates can also be part of a protein. The *quaternary structure* describes how these different chains and components interact and connect to each other by hydrogen bonding, electrostatic attraction and *sulfide bridges*. Such sulfide bridges are formed by *oxidation* of the –SH groups of Cysteine (Fig. 1.18). The product of this reaction is a covalently bonded dipeptide called *Cystin*.

The hormone insulin, which is produced in the pancreas, contains two different polypeptide chains, *A* and *B*. Sulfide bridges can occur within a chain as well as between the two chains (Fig. 1.19).

The separate polypeptide chains forming a protein can be identical (*homogenic* protein) or, as in the case of insulin, different (*heterogenic* protein).

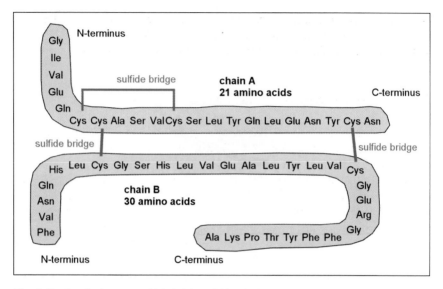

Fig. 1.19. Insulin has one sulfide bridge within chain *A* and two sulfide bridges between chain *A* and chain *B*.

1.1.2.3   *Degradation of proteins*

The three-dimensional structure of a protein which is held together by hydrogen bonding, electrostatic attraction and sulfide bridges is very sensitive to its chemical and physical environment. A change in pH, temperature or ionic strength disrupts these interactions and causes the protein to unfold; this process is called *denaturation*. The protein loses activity once its normal shape is lost. In some cases, this denaturation is reversible and the protein can *renaturate*, although in most cases the activity loss is permanent.

1.2   Nucleic Acids

Nucleic acids are long, linear biomolecules that can have molecular weights of several million Da. There are two classes of nucleic acids, *deoxyribonucleic acid (DNA)* and *ribonucleic acid (RNA)*.

DNA contains the "code of life." It is the hereditary molecule in all cellular life forms as it is used by cells to store and transmit genetic information. During cell division, exact copies of DNA are made. Cells use DNA to determine and control the synthesis of proteins with the help of messenger RNA (mRNA).

RNA is essential for the synthesis of proteins in the cells. Messenger RNA (mRNA) is synthesised in the cell nucleus as a transcript of a specific part of DNA.

The mRNA leaves the nucleus and enters the cell cytoplasm where it dictates the synthesis of proteins from amino acids. *Transfer RNA* (tRNA) delivers amino acids to the exact place in the cytoplasm where the proteins are synthesised.

1.2.1   *The Structure of Nucleic Acids*

Nucleic acids are made up form three components: *nucleobases* (usually referred to as bases), *sugars* and *phosphoric acid*. The nucleobases are derivatives of purine and pyrimidine (Figs. 1.20 and 1.21). Both DNA and RNA contain the purines Adenine (A) and Guanine (G). Of the pyrimidines, Thymine (T) and Cytosine (C) are components of DNA whereas Uracil (U) and Cytosine (C) are components of RNA. The sugar component of DNA is β-D-deoxyribose, while RNA contains β-D-ribose, (Fig. 1.22). These components are summarised in Table 1.4.

How are these linked to each other to form macromolecular DNA and RNA? A *nucleoside* is formed by one of the nucleobases covalently binding to a sugar (Fig. 1.23, left). This N-glycosidic bond is formed between the C1$'$ atom of the

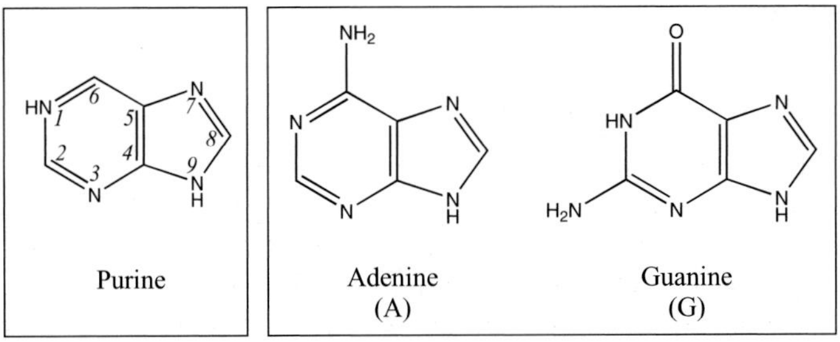

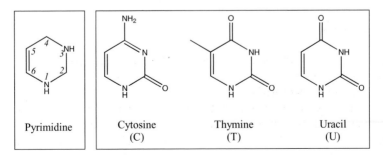

Fig. 1.20. Left: The structure of Purine, Right: The Purine derivatives Adenine and Guanine are found as bases in both DNA and RNA.

Fig. 1.21. Left: The structure of Pyrimidine. Right: Cytosine and Thymine are the Pyrimidine derivatives found in DNA, while Uracil and Thymine are found in RNA.

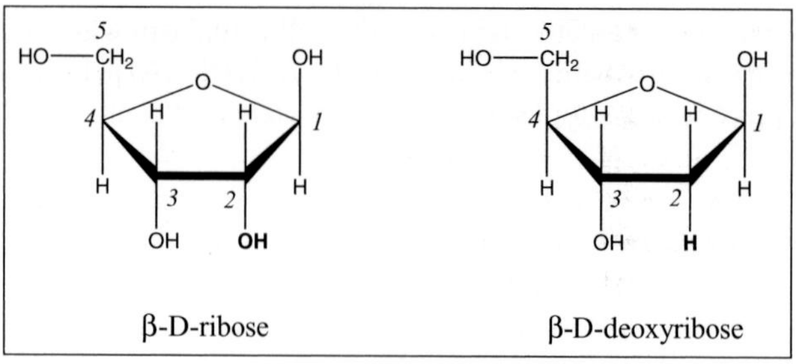

Bioanalytical Chemistry

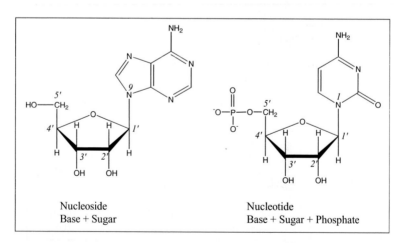

β-D-ribose   β-D-deoxyribose

Fig. 1.22. The pentose β-D-ribose occurs in RNA. β-D-deoxyribose is the sugar component in DNA.

Table 1.4. Components of nucleic acids.

		DNA	RNA
Nucleobases	**Purines**	Adenine Guanine	Adenine Guanine
	Pyrimidines	Thymine Cytosine	Uracil Cytosine
Sugar		β-D-deoxyribose	β-D-ribose
Phosphate		phosphoric acid	phosphoric acid

Nucleoside
Base + Sugar

Nucleotide
Base + Sugar + Phosphate

Fig. 1.23. Left: A nucleoside consisting of the base Adenine (A) and ribose. Right: A nucleotide consisting of Cytosine, ribose and one phosphate group.

sugar and either the N9 atom of a purine base or the N1 atom of a pyrimide base. A *nucleotide* is a nucleoside with one or more phosphate groups covalently attached to the 5′-hydroxylic group of the sugar (Fig. 1.23, right). At pH 7, the acidic phosphate groups are negatively charged.

Nucleotide units can connect to each other to form a chain, a *polynucleotide*. The phosphate residue at the 5′ position of one sugar bonds to the 3′-hydroxyl group of another sugar (Fig. 1.24). Thus, a strand is formed with the sugar and

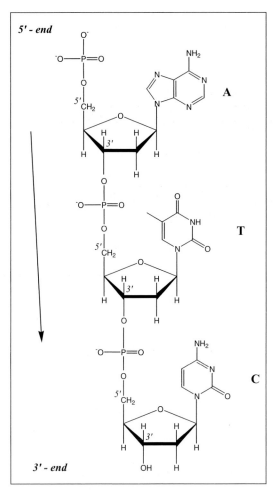

Fig. 1.24. Left: structure of a polynucleotide with the bases Adenine (A), Thymine (T) and Cytosine (C). Right: Abbreviated writing of the same polynucleotide sequence, as bar structure and as listing of nucleobases with hyphens.

phosphate units as a *backbone* and the nucleobases as *side groups*. In higher life forms, these strands can consist of millions of nucleotides.

To give the sequence a start and an end point, it is read from the nucleotide with the free phosphate group (the 5′-end) to the nucleotide with the free 3′-hydroxyl-group (the 3′-end).

The chemical structure of the polynucleotide can be described as shown in the left hand side of Fig. 1.24 or by using short forms with bars or by just listing the nucleobases in the sequence and omitting the sugar and phosphate.

1.2.1.1 *3D structure of DNA*

The three-dimensional structure of DNA was discovered by Francis Crick and James Watson in 1953. DNA as found in the cell nucleus has the shape of a *right twisted double helix* consisting of two polynucleotide strands twisted around each other (Fig. 1.25, left). The hydrophilic backbones composed of the sugar and phosphate groups are on the outside of the helix, while the hydrophobic bases are on the inside. The bases are connected to each other by weak hydrogen bonds to form *base pairs*. (Fig. 1.25, middle and right). The two strands run in opposite directions so that the 3′-end of one strand and the 5′-end of the other strand are linked.

Due to steric reasons, only two combinations of base pairs are possible: *Adenine with Thymine (A-T or T-A)* and *Guanine with Cytosine (G-C or C-G)* (Fig. 1.26).

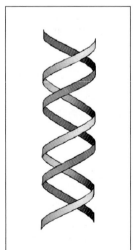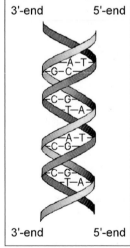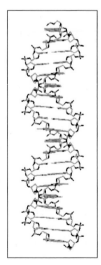

Fig. 1.25. Left: Schematic of a double helix. Middle: A DNA double helix with base pairs and sequence direction. Right: Structure of a DNA helix showing the parallel base pairs inside the helix.

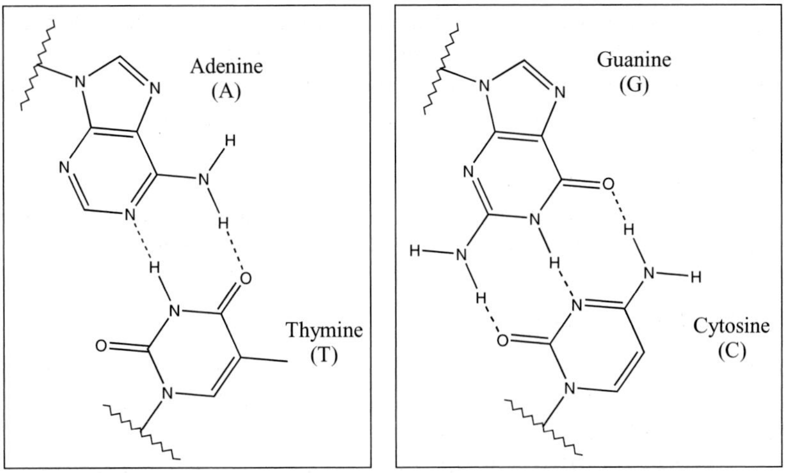

Fig. 1.26. Base pairs are formed between Adenine and Thymine (A-T) and Guanine and Cytosine (G-C).

The aromatic rings of the base pairs are parallel to each other, forming a twisted ladder-like structure. The diameter of a DNA helix molecule is about 2 nm. With millions of nucleotides in the strand, the length of a DNA molecule when laid out straight can measure several centimetres.

The two strands of the double helix are *complementary* to each other. The nucleobase-sequence of one strand unambiguously determines the sequence of the other strand.

Similar to the three-dimensional structure of proteins, three levels of organisation can be distinguished for DNA. The *primary structure* is determined by the sequence of nucleotides, usually written as the sequence of bases they contain. The *secondary structure* is given by the shape of the double stranded helix. This helical chain does not exist as a straight, long molecule. It forms turns and twists and folds. This coiling is referred to as the *tertiary structure* of DNA.

In comparison to proteins, DNA is a fairly simple molecule. It consists of only four different bases, which are repeated throughout the whole structure and the double helix is its only structural component.

When heated e.g. to 95°C or when deviating from physiological conditions, the hydrogen bonds between the two DNA strands are cleaved and the strands are separated from each other to form *single stranded DNA (ssDNA)*. This *denaturation* is usually a reversible process. When reverting to lower temperatures or to physiological conditions, the two strands can link back together to reform the double helix. Denaturation into ssDNA is a necessary step for the replication of DNA. Once cleaved, complementary daughter strands can be formed, which are an exact copy of the original strand. Denaturation into ssDNA is also essential for the synthesis of mRNA in cell nuclei and for the polymerase chain reaction (PCR), which is used for DNA amplification (see section 6.2).

1.2.1.2   *3D-structure of RNA*

RNA is also a polynucleotide, but with ribose as the sugar component and Uracil (U) as a base instead of Thymine (T). Due to the additional –OH group on the ribose sugar molecule, steric hindrance is too great to allow for the formation of a double strand. Hence, RNA can only exist as a *single stranded* molecule. This strand can fold and loop and form base pairing with itself in certain places.

1.2.2   *Synthesis of Proteins*

One of the main tasks of the DNA is to initiate the synthesis of proteins as and when they are needed. Proteins are synthesised in the ribosomes of the cell cytoplasm. DNA, however, is found in the cell nucleus. So how is the information contained in the DNA passed out of the cell nucleus and into the cytoplasm? First, the DNA helix unfolds, and, in a process called *transcription*, a complementary strand of RNA is synthesised along a crucial part of one of the single DNA strands. This is the *messenger RNA* (*mRNA*) which leaves the cell nucleus and is transported into the manufacturing centres for proteins, the ribosomes. In the ribosome, *transfer RNA* (*tRNA*) delivers the amino acids required for polypeptide synthesis. The sequence of each group of three bases on the mRNA determines which amino acid is next in the peptide sequence. For example, the sequence AGC in the mRNA specifies the incorporation of the amino acid serine. This process is referred to as *translation* (Fig. 1.27). The genetic code, i.e. which sequence of bases in the DNA strand refers to which amino acid is given in Table 1.5.

To obtain its biological activity, the synthesised protein must fold into its native structure. Disulfide bridges must be formed. If the protein has a quaternary structure, then the different peptide chains must combine. Often, the protein undergoes a number of *post-translational modifications* to gain its full activity. The most common post-translational modification is the cleavage of one or several amino groups from the N- or C-terminus of the peptide chain. The side chains of the amino acids can undergo chemical modifications such as phosphorylation, acetylation, and methylation. *Glycoproteins* are synthesised by glysosilation, i.e. the addition of an oligosaccharide to the peptide chain. Similarly, *lipoproteins* can be formed. More than 150 such post-translational modifications are known. None of these is determined by the DNA sequence. However, these modifications are crucial for the biological activity of the proteins.

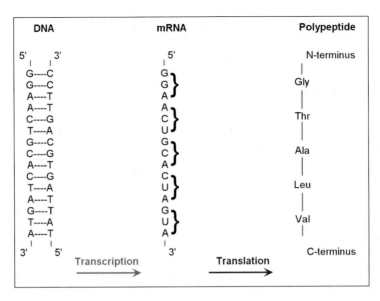

Fig. 1.27. Transcription of a DNA sequence into mRNA followed by translation into a polypeptide sequence.

Table 1.5. Genetic code. Each sequence of three bases in the mRNA determines which amino acid is used in the polypeptide (refer to Table 1.1 for amino acid abbreviations).

First position 5′ end	Second position				Third position 3′ end
	U	C	A	G	
U	UUU Phe	UCU Ser	UAU Tyr	UGU Cys	U
	UUC Phe	UCC Ser	UAC Tyr	UGC Cys	C
	UUA Leu	UCA Ser	UAA Stop	UGA Stop	A
	UUG Leu	UCG Ser	UAG Stop	UGG Trp	G
U	CUU Leu	CCU Pro	CAU His	CGU Arg	U
	CUC Leu	CCC Pro	CAC His	CGC Arg	C
	CUA Leu	CCA Pro	CAA Gln	CGA Arg	A
	CUG Leu	CCG Pro	CAG Gln	CGG Arg	G
A	AUU Ile	ACU Thr	AAU Asn	AGU Ser	U
	AUC Ile	ACC Thr	AAC Asn	AGC Ser	C
	AUA Ile	ACA Thr	AAA Lys	AGA Arg	A
	AUG Met	ACG Thr	AAG Lys	AGG Arg	G
G	GU Val	GCU Ala	GAU Asp	GGU Gly	U
	GUC Val	GCC Ala	GAC Asp	GGC Gly	C
	GUA Val	GCA Ala	GAA Glu	GGA Gly	A
	GUG Val	GCG Ala	GAG Glu	GGG Gly	G

1.3 Biomolecules in Analytical Chemistry

Chemical analysis of the biomolecules introduced in the preceding sections is often radically different from the "classical" analysis of relatively small organic molecules such as pesticides, drugs or explosives residues.

The analytical chemist is faced with a variety of tasks, which can be roughly separated into four categories: (1) qualitative analysis of a mixture of compounds, (2) qualitative analysis of a pure compound, (3) quantitative analysis of a selected compound in a mixture and (4) structure elucidation of a pure compound.

1.3.1 *Classical Analytical Chemistry*

A number of methods are used in classical analysis to perform these tasks. Qualitative as well as quantitative analysis of mixtures can be achieved by chromatographic methods such as gas chromatography (GC) and liquid chromatography (LC). Chemical sensors or biosensors can also be employed for selectively quantifying a compound in a mixture. However, such analysers have only been developed for a very limited number of analytes. Identification of pure compounds can be achieved by nuclear magnetic resonance (NMR) measurements, by mass spectrometry (MS), infrared spectroscopy (IR), UV/vis spectroscopy or X-ray crystallography, to name a few.

1.3.2 *Limitations of Classical Analytical Chemistry*

Most of these "classical" methods are not, or only to a limited extent, suitable for the analysis of nucleic acids and proteins. GC requires the analytes to be volatile and thermally stable, a property rarely exhibited by a biomolecule. MS is often used for molecular weight determination. However, the high molecular weight biomolecules are extensively fragmented by conventional ionisation methods, making molecular weight (MW) analysis impossible. NMR-spectroscopy works extremely well for structural elucidation of medium sized molecules. However, the sheer number of spin-active nuclei, especially ^{1}H, in any single DNA or protein molecule results in an enormous number of signals that make structural elucidation extremely difficult if not impossible. A similar problem holds true for structural elucidation via IR- and UV-spectroscopy. The number of functional groups and chromophores makes unambiguous identification impossible.

Biochemists are often interested in parameters that are not applicable for most classical analytes. Nucleic acids and proteins are both large biopolymers, consisting of sometimes thousands of monomers linked together. Methods are required to accurately determine the *sequence* of amino acids in a protein or the sequence of bases in a DNA or RNA molecule. Changing a single amino acid in the protein might alter its folding pattern and biological activity. Changing a single base in a DNA strand might cause a genetic disorder. Not only must the sequencing methods be accurate, they should also have potential for automation, if large amounts of samples are to be analysed and compared. *Separation* methods for biomolecules must be extremely powerful. The number of proteins in a single cell can run into thousands. For sequencing, the biomolecules are often partially digested into smaller fragments, which have to be separated from each other. Separation methods are needed on both, preparative and analytical scales. The amount of sample available is often very small. Hence, high sensitivity is required. Very sophisticated methods are needed to elucidate the three-dimensional structure of a biomolecule with a molecular weight of possibly thousands of kDa.

1.3.3   *Bioanalytical Chemistry*

Without the development and improvement of bioanalytical methods over the recent decades, the enormous progress in genomics and proteomics would have been impossible. Methods for accurate determination of high molecular weights, for sequencing of DNA and proteins and for separating thousands of molecules in a single run have revolutionised analytical chemistry. Many of these methods have been transformed into commercially available bench top instruments that offer high-throughput, automated and computer controlled analyses. In Table 1.6, a number of methods employed for nucleic acid and protein analysis are summarised. The list is by no means exhaustive. It is intended to give an overview and to emphasise the difference between classical analysis and bioanalysis.

1.3.3.1   *Analysis of nucleic acids*

The field of *genomics* concerns the study of the entire genome of a cell or an organism, i.e. the complete DNA sequence, and the determination of all the genes within that sequence. The human genome consists of 3 billion base pairs. An estimated 30,000-40,000 genes are contained within this sequence. Scientists are trying to map the locations of genes to improve our understanding of genetic disorders, and to explore the organisation and interplay of these genes.

Before any sequencing reactions can be carried out, the nucleic acids must be isolated from the cell and purified. To obtain a sufficient amount of sample, amplification of the DNA molecules is usually required. Analysis of nucleic acids can, thus, be divided into the following steps: (1) isolation and purification, (2) quantification and amplification and (3) sequencing.

Table 1.6. Comparison of classical and bioanalytical chemistry.

Analytical task	Classical analytical chemistry	Bioanalytical chemistry	
	Small molecules	DNA, RNA	Proteins
investigate a mixture qualitatively	GC	CE	MALDI-TOF-MS
	LC	GE, 2D-GE	
selectively quantify a compound in a mixture	GC	CE	bioassay
	LC	Real-Time PCR	
	chemical or biosensor	DNA array	
		biosensor	
identify a pure compound qualitatively	MS	PCR	amino acid composition
	NMR	DNA arrays	tryptic digest and GE MALDI-TOF-MS ESI-MS
elucidate the structure of a pure compound	NMR	DNA sequencing	amino acid sequencing
	MS	NMR	
	IR	X-ray crystallography	
	X-ray crystallography	electron-microscopy	

To isolate the DNA or RNA molecules from the cell, a number of conventional methods such as liquid-liquid extraction, precipitation and centrifugation (section 6.1) can be employed.

An isolated nucleic acid can then be quantified, for example by UV spectroscopy. The aromatic groups of the bases have an absorption maximum around $\lambda = 260$ nm. Alternatively, fluorescent or radioactive markers can be attached and quantitatively detected. A mixture of DNA molecules can be quantified by capillary electrophoretic methods (section 3.3).

A DNA molecule can be amplified by the polymerase chain reaction (PCR) (section 6.2), if part of its sequence is known. One DNA molecule is sufficient to generate millions of identical copies in a controlled amplification reaction. With real-time PCR, the DNA quantity can be measured during the amplification reaction (section 6.2.4). Other methods of DNA quantification include DNA arrays (section 5.3) and, if available, biosensors (section 5.2).

Sequencing methods for DNA include the Maxam-Gilbert method (section 6.3.2), the Sanger method (section 6.2.4), DNA arrays (section 5.3) and pyrosequencing (section 5.4). Usually, the DNA molecules are treated with a restriction enzyme (section 6.3.1) prior to sequencing. A number of fragments are thus generated, which are then separated from each other according to their molecular weight by gel electrophoresis (sections 3.2 and 3.3.5). The most efficient separation method is two-dimensional gel electrophoresis (2D-GE) (section 3.2.5).

1.3.3.2 *Analysis of proteins*

In *proteomics* research, the aim is to study all the proteins expressed in a cell, tissue or organisms to obtain an insight into the interplay of cells and organism. Protein analysis often involves isolation and investigation of one protein at a time.

Protein concentrations in cells are usually very low and an amplification reaction such as PCR for DNA molecules does not exist for proteins. Isolating a protein from a complex cell matrix with a high yield and without changing its biological functionality can be a difficult task. Some of the analytical methods involved are liquid-liquid extraction, precipitation and centrifugation. Often a protein or a group of proteins is separated from impurities by liquid chromatography (section 2) or polyacrylamide gel electrophoresis (PAGE) (section 3.2).

Quantitative analysis of proteins can be achieved by UV spectroscopy. The peptide bond has an absorption maximum around $\lambda = 205$ nm, the aromatic rings on the amino acids Tryptophan and Tyrosine absorb strongly around $\lambda = 280$ nm. Also commonly used are colorimetric assays, which contain reagents that specifically form coloured complexes with proteins. These quantitative methods usually measure the total protein concentration. Either the protein of interest has to be isolated prior to analysis, or a very specific method has to be found to quantify only the targeted protein. Very sensitive and specific analysis of antibodies and antigens can be achieved with bioassays (section 5.1) or biosensors (section 5.2).

The amino acid composition (section 7.5) of a protein can be determined by first completely hydrolysing the peptide bonds and then separating and quantifying the obtained amino acids. Ion exchange chromatography (IEC) (section 2.3.2), reversed phase liquid chromatography (RP-HPLC) (section 2.3.1) and capillary electrophoresis (CE) (section 3.3) can be employed for separation and quantification. If the protein is known and has been catalogued, the amino acid composition is often enough to unambiguously identify a protein in a sample.

Alternatively, the protein can be partially digested by an enzyme like trypsin. The fragments of this tryptic digest can then be separated and their molecular weights can be measured with a mass spectrometer. Methods available for such analysis are matrix assisted laser desorption ionisation (MALDI) time of flight (TOF) mass spectrometry (MS) (section 4.1) as well as electrospray ionisation mass spectrometry (ESI-MS) (section 4.2) coupled to liquid chromatography (LC-ESI-MS) or capillary electrophoresis (CE-ESI-MS). Often the protein can be identified from the molecular weights of the tryptic digest fragments.

If the protein cannot be determined by either of these two methods, then sequencing of the amino acids becomes necessary. The strategies for this are outlined in chapter 7.

For proteomics analysis, all the proteins from a cell must be extracted and then separated from each other. Gel electrophoretic methods (section 3.2) are most powerful, especially two-dimensional gel electrophoresis (2D-GE), which is capable of separating thousands of proteins in a single run (section 3.2.5).

Three-dimensional structures of both proteins and nucleic acids can be obtained by sophisticated NMR experiments, by electron microscopy, and by X-ray crystallography, if a monocrystal can be obtained. Covering these techniques is beyond the scope of this book. The reader may refer to one of the textbooks in the references given at the end of this chapter.

Summary

The structure and main features of amino acids, proteins nucleotides and DNA were outlined in this chapter.

DNA is the hereditary molecule of all cellular life forms. It stores and transmits genetic information. DNA is a relatively simple molecule, composed only of four different nucleotides with the bases adenine, guanine, thymine and cytosine and β-D-deoxyribose as the sugar component. Millions of nucleotides can be linked together. Two complementary strands are twisted around each other in the form of a double helix. They are held together by hydrogen bonds between the base pairs adenine-tymine and cytosine-guanine. RNA is comprised of nucleotides with the bases adenine, guanine, uracil and cytosine and β-D-ribose as the sugar component. RNA is a single stranded molecule with base pairing occurring only in parts of this single chain.

Proteins are relatively complicated molecules made up from the 20 naturally occurring α-L-amino acids, which are linked to each other via peptide bonds. Fibrous proteins give mechanical strength to bones and muscles. Globular proteins such as antibodies and enzymes have specific functions in the immune system and in metabolism. The macromolecular chain of a protein is folded in a very specific way. This folding is essential for the protein's function and activity. The sequence of the amino acids is referred to as the primary structure of the protein. Parts of the amino acid chain form domains of regular structures such as α-helices and β-pleated sheets. These make up the secondary structure of the protein. The whole three-dimensional shape of the amino acid chain including interactions between different secondary domains is referred as the tertiary structure. Some proteins consist of two or more chains and other, non-amino acid components. This is described as the quaternary structure of the protein.

The three dimensional structure of both, DNA and proteins, is only stable within a certain chemical and physical environment. Changes of temperature, pH and ionic strength may cause the biomolecule to denature. This denaturation is often irreversible.

In a cell, the DNA contains the genetic code. When a gene is "switched on", it triggers the synthesis of a protein. This protein synthesis is achieved by the processes of transcription and translation.

The methods for analysing and identifying biomolecules are radically different from analysing relatively small organic molecules. Separation of biomolecules is commonly carried out by gel and capillary electrophoresis. Chromatography is used not so much as a separation method, but mainly as a method for purification and isolation of compounds. The molecular recognition that many biomolecules exhibit is used in many analytical tools including immunoassays, biosensors and DNA arrays. The structure of a biomolecule cannot easily be determined by spectroscopic methods. Determination of a protein as well as a nucleic acid structure involves a number of reactions and analysis steps to be carried out.

References

1. L. Stryer, J. M. Berg and J. L. Tymoczko, *Biochemistry*, 5th edition, W. H. Freeman and Co., 2002.
2. D. L. Nelson, A. L. Lehninger and M. M. Cox, *Principles of Biochemistry*, 3rd edition, Worth Publishers, 2000.
3. B. Alberts, A. Johnson, J. Lewis, M. Raff, K. Roberts and P. Walter, *Molecular Biology of the Cell*, 4th edition, Garland Science Publishing, 2002.
4. H. Lodish, A. Berk, L. S. Zipursky, P. Matsudaira, D. Baltimore and James Darnell, *Molecular Cell Biology*, 4th edition, W. H. Freeman, 2000.

Chapter 2

CHROMATOGRAPHY

In this chapter, you will learn about...
♦ ...the principles of separation in chromatography.
♦ ...the basic separation theory.
♦ ...chromatographic methods which are commonly applied to the separation of biomolecules.

Chromatography is used routinely in almost every (bio)chemical laboratory for a large number of tasks. These range from the *separation* of mixtures on an analytical as well as preparative scale, from *purification* and preconcentration of an analyte, to *controlling* the progress of a chemical reaction. Since the first description of chromatography by Russian botanical scientist Mikhail Semenovich Tswett in the early 20th century, an enormous variety of formats and applications has been developed. To describe all of them in detail would be beyond the scope of this textbook. The focus of this chapter will concern the employment of chromatography for the separation and purification of biomolecules. These will be outlined after a brief explanation of the *principles* of chromatographic separation and a short summary of the basic *equations* used in chromatographic theory.

2.1   The Principle of Chromatography

Chromatography is a separation method where the analyte is contained within a liquid or gaseous *mobile phase*, which is pumped through a *stationary phase*. Usually, one phase is hydrophilic and the other lipophilic. The components of the analyte interact differently with these two phases. Depending on their polarity, they spend more or less time interacting with the stationary phase and are thus retarded to a greater or lesser extend. This leads to the separation of the different components present in the sample. Each sample component *elutes* from the stationary phase at a specific time, its *retention time* t_R (Fig. 2.1). As the components pass through the detector, their signal is recorded and plotted in the form of a *chromatogram*.

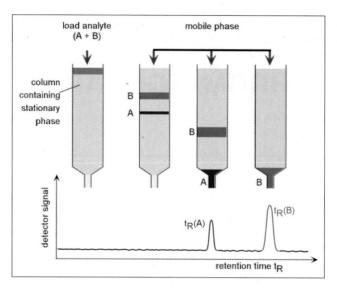

Fig. 2.1. The principle of chromatographic separation. The sample components interact differently with the stationary and mobile phase and elute at their specific retention time, t_R.

Chromatographic methods can be classified into gas chromatography (GC) and liquid chromatography (LC) depending on the nature of the mobile phase involved.

Gas chromatography can be applied only to gaseous or volatile substances that are heat-stable. The mobile phase, an inert carrier gas such as nitrogen, hydrogen or helium, is pumped through a heated column. This column can be packed with a silicon oxide based material or is coated with a polymeric wax. The sample is vaporised, pumped through the column and the analytes are detected in the gas stream as they exit the column. Analyte detection can be achieved by either flame ionisation or thermal conductivity. GC is not commonly used for the analysis of biomolecules since large molecular weight compounds such as peptides and proteins are thermally destroyed before evaporation. Smaller molecules such as amino acids, fatty acids, peptides and certain carbohydrates can be analysed if they are modified chemically to increase their volatility. Some cell cultures produce volatile metabolites such as aldehydes, alcohols or ketones. These can be analysed readily via GC.

In *liquid chromatography*, the sample is dissolved and pumped through a column containing the stationary phase. LC is more versatile than GC as it is not restricted to volatile and heat-stable samples; the sample only has to dissolve completely in the mobile phase. Common detection methods are UV spectroscopy, measurement of refractive index, fluorescence, electrical conductivity and mass spectrometry. Modes of operation can be classified as normal and reversed phase chromatography. In *normal phase* chromatography, the stationary phase consists of a hydrophilic material such as silica particles and the mobile phase is a hydrophobic organic solvent such as hexane. In *reversed phase* chromatography, on the other hand, the stationary phase is hydrophobic and the mobile phase is a mixture of polar solvents, for example water and acetonitrile. Biomolecules are generally soluble in polar solvents, hence, reversed phase chromatography is the method of choice for amino acids, peptides, proteins, nucleic acids and carbohydrates.

2.2   Basic Chromatographic Theory

The optimisation of chromatography is aimed towards completely separating all of the components of a sample in the shortest possible time. This can, for example be achieved by modifying the composition of the mobile phase, choosing a different stationary phase or by changing the flow rate. A typical chromatogram is depicted in Fig. 2.2. The sample is injected into the chromatographic column at $t = 0$ s. Substances that are not retarded by the stationary phase leave the column at *zero retention time*, t_0, corresponding to the flow rate of the mobile phase. Compounds A and B are retarded by the stationary phase and leave the column at their retention times $t_R(A)$ and $t_R(B)$, respectively. The peak width, w, is defined as the intersection of the tangents on each side of the peak with the baseline.

These basic parameters, retention time and peak width, can be used to derive a number of other parameters that express the quality of the achieved chromatographic separation. In the following paragraphs, a brief summary of the most important parameters of chromatographic theory are discussed.

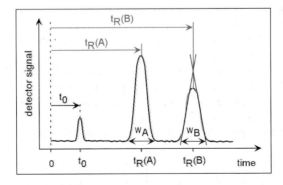

Fig. 2.2. Definition of retention time, t_R, and peak width, w.

The *capacity factor k'* (equation 2.1) describes the velocity of the analyte relative to the velocity of the mobile phase. Each compound spends a different amount of time interacting with the mobile and stationary phase. The average velocity of a sample compound is dependent on how much time it spends in the mobile phase. If k' is much smaller than 1, then the analyte moves too quickly and the elution time is so short that an exact determination of t_R is difficult. If the sample moves too slowly, the separation time is very high. A good value for k' would be between about 1 and 5. The *selectivity factor α* (equation 2.2) describes the relative velocities of the analytes with respect to each other. The selectivity describes how well a chromatographic method can distinguish between two analytes.

capacity factor $\qquad k' = \dfrac{t_R - t_0}{t_0}$ $\qquad\qquad$ (equation 2.1)

selectivity factor $\qquad \alpha = \dfrac{k'_B}{k'_A} = \dfrac{t_R(B) - t_0}{t_R(A) - t_0}$ $\qquad$ (equation 2.2)

The *efficiency* of a chromatographic separation is crucially dependant on band broadening. If band broadening is large, peaks can overlap and resolution is lost. Band broadening for a column of length L is quantitatively expressed in the concept of *height equivalent of a theoretical plate, H, or* simply *plate numbers, N* (equations 2.3 and 2.4). The larger the number of plates N and the smaller H is, the better the chromatographic efficiency.

plate number $\quad N = 16 \left(\dfrac{t_R}{w} \right)^2$ $\qquad\qquad$ (equation 2.3)

plate height $\quad H = \dfrac{L}{N}$ $\qquad\qquad$ (equation 2.4)

The parameters that influence band broadening can be approximated by the *van Deemter equation* (equation 2.5) which is valid for gas and liquid chromatography as well as capillary electrophoresis (see chapter 3.2).

van Deemter equation $H = A + \dfrac{B}{u} + C \cdot u$ $\qquad$ (equation 2.5)

In this simplified equation, the height of theoretical plates, H, is given as a sum of three terms. The first term, A, describes the influence of the column packing on band broadening. This so-called *Eddy diffusion* is constant for a given column and independent of the flow rate. The second term, B/u, describes the diffusion in or opposed to the direction of flow. This *longitudinal diffusion* is inversely proportional to the flow rate u. The third term, $C \cdot u$, describes the resistance to *mass transfer* between the stationary and mobile phase which is directly proportional to the flow rate. By plotting H as a function of u, the optimum flow rate for a chromatographic separation can be determined (Fig. 2.3).

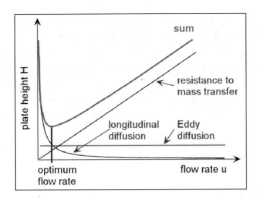

Fig. 2.3. A van Deemter plot for the determination of the optimum flow rate.

The ultimate goal of a separation is to achieve a high resolution, R_s, (equations 2.6 and 2.7). If $R_s = 1.5$, then peaks of identical area overlap by only 0.3 %, an $R_s = 1$ equals a peak overlap of 4 %. Peak resolution can be optimised by increasing the selectivity and minimising band broadening.

$$\text{resolution} \qquad R_s = \frac{2 \cdot [t_R(A) - t_R(B)]}{w_A + w_B} \qquad \text{(equation 2.6)}$$

$$R_S = \frac{\sqrt{N}}{4}(\alpha - 1)\left[\frac{k'}{1+k'}\right] \qquad \text{(equation 2.7)}$$

(valid for $\alpha < 1.2$)

As can be seen from equation 2.7, the capacity factor k' has a great influence on the resolution. Usually the components in the sample have a wide variety of k' values. If conditions are optimised such that the first compounds to elute have k' values between the optimum of 1 and 5, then the other compounds with higher k' values elute much later and show excessive band broadening. If, on the other hand, conditions are optimised for the later eluting compounds, then the resolution will be poor for the compounds that elute first. This *general elution problem* can be overcome by decreasing k' during the separation. In LC, the composition of the mobile phase can be changed during the separation. This is called a *gradient elution* as opposed to an *isocratic elution*, where the composition of the mobile phase remains unchanged during the separation process. In GC, a temperature gradient can be applied during separation rather than operating under isothermic conditions.

Generally, the first step in trying to achieve a good separation of the sample mixture is to choose a stationary phase with which the analyte can interact. Then, the composition and gradient of the mobile phase can be chosen to optimise the capacity factor and resolution.

Chromatographic theory as outlined in the above paragraphs can be applied to the analysis of smaller molecules such as amino acids, peptides and short biopolymers. Care has to be taken for larger biomolecules such as high molecular weight proteins. These often show different behaviour and the theory can only be applied to a limited extent.

2.3   Application of Liquid Chromatography for Bioanalysis

In bioanalytical chemistry, chromatography is mainly used for the separation, isolation and purification of proteins from complex sample matrices. In cells, for example, proteins occur alongside numerous other compounds such as lipids and nucleic acids. In order to be analysed, these proteins must be separated from all the other cell components. Then the protein of interest might have to be isolated from other proteins and purified further. Chromatography is an essential part of almost any protein purification strategy. A number of different chromatographic techniques are used for the purification and analysis of proteins. They can be classified according to the physical principle involved in the separation process. Typical examples include *reversed phase chromatography, ion exchange chromatography, affinity chromatography* and *size exclusion chromatography (SEC)* (Table 2.1). These are outlined in more detail in the following sections.

Table 2.1. Separation methods for proteins.

Property	Separation method
hydrophobicity	reversed phase chromatography
charge	ion exchange chromatography
biospecificity	affinity chromatography
size, form	size exclusion chromatography

2.3.1   *Reversed Phase Liquid Chromatography*

Normal phase chromatography was developed many years before reversed phase chromatography was investigated. Initially, *stationary* phases were made of *polar* materials such as paper, cellulose or silica gel and the *mobile phase* consisted of *non-polar* solvents such as hexane or chloroform. Only at a later stage were these phase polarities *reversed. Polar solvents* such as water and acetonitrile were

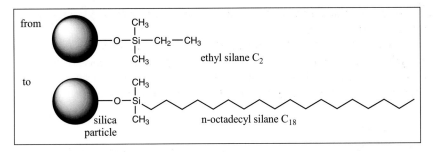

Fig. 2.4. Surface groups used for stationary phases in reversed phase chromatography range from ethyl silane (C_2) to n-octadecyl silane (C_{18}).

used in combination with *non-polar stationary* phases. These were obtained by etherification of the polar hydroxyl groups of the silica gel with long alkyl chains.

Reversed phase chromatography is the method of choice for the separation of smaller biomolecules such as peptides, amino acids, carbohydrates and steroids, which are soluble in water/acetonitrile mixtures. The separation of proteins can be problematic as organic solvents such as acetonitrile can decrease the protein's solubility and cause denaturisation.

The *stationary phase* usually consists of porous silica particles with *non-polar* surface groups (Fig. 2.4), obtained from etherification of the initial hydroxyl groups of the silica particle with *silanes* containing non-polar hydrocarbon chains. Any chain length from ethyl silane (C_2) to n-octadecyl silane (ODS) (C_{18}) is used, although *octyl silane* (C_8) and *ODS* are the most commonly employed chain lengths. For analytical separations, the particle size is typically $5\,\mu m$ or smaller. In preparative liquid chromatography, where the goal is to isolate a compound of interest for further analysis or investigation, larger particles with a higher capacity and larger column diameters are used. The pore size of the silica particles is usually about 10 nm, resulting in a very large surface area, as much as 100 to $400\,m^2g^{-1}$. This gives the analytes ample opportunity to interact with the stationary phase whilst flowing through the separation column.

The *mobile phase* is based on a *polar* solvent system consisting of an aqueous buffer and acetonitrile or methanol. Gradient elution is often employed to increase resolution and shorten separation times. This is achieved by increasing the organic solvent and thus decreasing the mobile phase polarity and the retention of less polar analytes during the separation process. Solvents can be classified according to their elution strength and polarity (Fig. 2.5).

Buffer systems based on ammonium acetate, phosphate or hydrogen carbonate are usually added at concentrations of about 20 mM to adjust the pH of the mobile phase to values between 2 and 8. *Ion pairing reagents* can be used at low concentrations, typically 0.1%, to increase the hydrophobicity of charged analytes. They

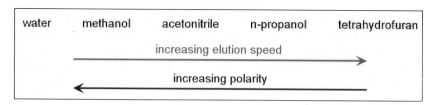

Fig. 2.5. Solvents ordered according to polarity and elution speed of the analytes.

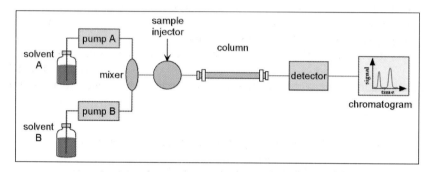

Fig. 2.6. Instrumental setup of an HPLC gradient system.

form ion-pair complexes with the analyte. Anionic ion pairing reagents such as trifluoroacetic acid (TFA) bind to positively charged analytes, whereas cationic ion pairing reagents such as tetraalkyl ammonium salts can be used to bind to negatively charged analytes. These complexes are retarded more by the stationary phase and are thus easier to separate than the largely unretained charged analytes alone.

In modern chromatography, the separation columns are tightly packed with small particles of about 1–5 μm in diameter. To achieve ambient flow rates in these columns, high pressures of up to 300–400 bar must be generated. A typical instrumental setup for this *high pressure* or *high performance liquid chromatography* (*HPLC*) is shown in Fig. 2.6.

Computer controlled *pumps* move the mobile phase through the system. Aqueous solvent *A* and organic solvent *B* are mixed to the desired composition. In the case of *gradient elution*, the composition is gradually altered during the separation. Sample volumes are injected with either a manual loop and valve system or automatically via an auto sampler. Depending on the column dimensions sample volumes can be as low as several nL and as high as a mL. Often the column is situated inside an oven which is thermostatically regulated to maintain a constant temperature. After eluting from the column, the analytes pass through the detector. *UV detection* using a fixed wavelength could be performed at $\lambda = 210$ nm for peptides and $\lambda = 254$ nm or $\lambda = 280$ nm for proteins (section 1.3). More expensive instruments have *diode array detectors* (DAD) which can take several whole spectra per second and allow for more unambiguous identification. High sensitivity can be achieved via *fluorescence detection* of derivatised amino acids and peptides. A more recent development is to couple liquid chromatography systems to an electrospray ionisation mass spectrometer, ESI-MS (section 4.2). Mass spectrometry allows universal detection at very high sensitivity and also gives structural information about the analyte. However, not all buffers commonly employed for liquid chromatography are compatible with mass spectrometers.

In recent years, there has been a trend to develop ever smaller liquid chromatography systems. LC systems on *micro* and even *nanoscales* have been demonstrated. Shorter and smaller columns with smaller particles offer faster analysis times, decreased solvent consumption and require less sample. The differences between preparative, analytical, micro and nano LC are summarised in Table 2.2.

Table 2.2. Differences between preparative, analytical, micro and nano liquid chromatography.

Method	Amount of sample	Column diameter in mm	Flow rate in $mL\ min^{-1}$
preparative	mg–g	>4	>1
analytical	μg–mg	2–4	0.2–1
micro	μg	1	0.05–0.1
nano	ng–μg	<1	<0.05

2.3.2 *Ion Exchange Chromatography*

Ion exchange chromatography separates and purifies analytes according to their overall charge. It can be used for almost any kind of charged molecule including large proteins, small nucleotides and amino acids. It is often used as a first step in protein purification. The principle of ion exchange chromatography is based on the competitive interaction between charged sample molecules and salt ions for the charged functional groups on the stationary phase.

This chromatographic process can be divided into several steps (Fig. 2.7). The column usually contains porous particles with positively charged functional groups on the surface. After adding the sample, negatively charged molecules bind to the surface groups on the stationary phase – they are *adsorbed*, whereas the neutral and positively charged sample components are not retained and elute from the

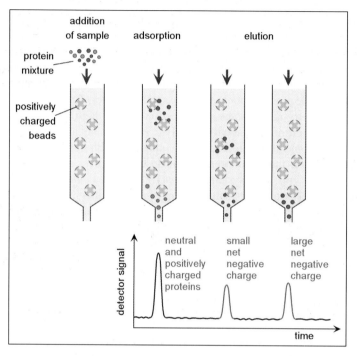

Fig. 2.7. Principle of ion exchange chromatography. Negatively charged sample components are adsorbed on the stationary phase and thus separated from positively charged and uncharged sample components. The adsorbed components are then eluted by increasing the ionic strength of the mobile phase.

column. The negatively charged analytes are then *desorbed* and *eluted* gradually by increasing the salt concentration of the mobile phase or by changing its pH.

The *stationary phase* used in ion exchange chromatography is often referred to as a gel. It consists of agarose or cellulose beads with covalently attached charged groups. Anion exchangers feature positively charged functional surface groups whereas cation exchangers feature negatively charged surface groups. Commonly used ion exchangers are *diethyl aminoethyl* (*DEAE*) and *carboxy methyl* (*CM*) (Table 2.3). It should be noted that the charge and thus the capacity of these ion exchangers depends on the pH of the mobile phase that is used. For example, an anion exchanger like DEAE will be deprotonated and thus neutralised at high pH and lose its activity. Both CM and DEAE work sufficiently well at pH values between 4 and 8, the range of greatest relevance for biomolecular applications.

Proteins are amphoteric molecules as they have basic amino groups and acidic carboxyl groups. The overall charge of a protein is the sum of the individual

Table 2.3. Functional groups for ion exchange stationary phases.

Name		Functional group	Type
diethyl aminoethyl	(DEAE)	$-(CH_2)_2-\overset{\oplus}{N}H(CH_2CH_3)_2$	anion exchanger
carboxy methyl	(CM)	$-CH_2-COO^{\ominus}$	cation exchanger

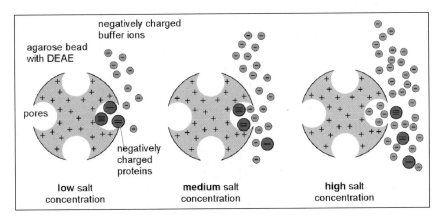

Fig. 2.8. Desorption of negatively charged analytes from an anion exchanger by increasing the salt concentration. Analytes with a low net charge are eluted at medium salt concentration, whereas analytes with a high net charge require a mobile phase with a high ionic strength before they are eluted.

charges of its component amino acids. Depending on the pH of the solvent, they either have a positive or a negative net charge. The pH at which the protein has no net charge defines the *isoelectric point, pI* (section 1.1.1.1). Adsorption of the protein to the stationary phase will be minimal when operating at a pH close to the pI. If, however, the pH differs considerably from the protein's pI, then the protein will have a high net charge and interact strongly with the charges of the stationary phase. To adsorb a protein onto a cation exchanger such as CM, the protein must be positively charged. Hence, the pH of the mobile phase must be lower than the protein's pI. On the other hand, to absorb a protein onto an anion exchanger such as DEAE, the protein must be negatively charged. Thus, the pH of the mobile phase must be adjusted so that it is higher than the pI of the protein. To minimise competition with buffer ions for the binding sites, buffer concentrations for the adsorption step are kept fairly low, between 10 and 20 mM. Commonly used buffers include phosphate and acetate salts.

Gradual desorption of the immobilised analytes is achieved either by a continuous *increase* in *ionic strength* or a *change* in *pH* of the mobile phase. Salt gradients, for example with NaCl, are often used for desorption on cation as well as anion exchangers. The salt concentration is gradually increased from 0 to 1 M or higher. The salt ions compete with the proteins for the charged binding sites (Fig. 2.8). Weakly charged proteins are eluted first, whereas strongly charged proteins are retained and elute only at very high ionic strengths. Above a certain ionic strength, all sample components become fully desorbed. This method allows bound proteins to be desorbed gradually, according to their net charge and to be eluted from the column consecutively. A change in pH can also be used for desorption. This can result in decreasing the net charge of the proteins or neutralising the functional groups of the ion exchanger. The interaction between the analyte and exchanger is weakened and, thus, the analyte is desorbed.

2.3.3   *Affinity Chromatography*

Affinity chromatography makes use of the highly specific *molecular recognition* of certain biomolecules. By attaching a specific ligand such as an antigen to the stationary phase material, the matching antibody can be specifically and reversibly adsorbed. Molecular recognition does not only occur between antigens and antibodies. Other bonding partners exist, including enzyme and co-enzyme, receptor protein and hormone or single strands of oligonucleotides and their matching counterparts.

Affinity chromatography has the highest specificity and selectivity of all chromatographic methods and is a powerful method for the purification and isolation of biomolecules even at low concentrations. The target molecule can be picked selectively from complex mixtures such as blood or serum.

The process can be divided into the following steps: (1) sample introduction, (2) adsorption, (3) washing and (4) desorption (Fig. 2.9).

The chromatographic column contains agarose or cellulose beads as a stationary phase on which ligand molecules have been covalently attached. After addition of the crude sample, those molecules that have an affinity for the ligand on the beads are *adsorbed* and retained by the stationary phase. Other substances of the sample mixture with no affinity for the ligand are eluted from the column (Fig. 2.10, left). Further washing ensures removal of non-specifically bound components. In the next step, the adsorbed species have to be *eluted* from the column. To achieve desorption, the non-covalent interaction between the biomolecules must be disrupted. Methods include a pH decrease, an increase in ionic strength (to several M), addition of a denaturing agent such as urea (to several M) or the addition of an organic solvent. These desorption methods are *non-specific* as they elute any bound molecule alike. *Specific desorption* can be achieved by introducing a species that binds to the analyte more strongly than the ligand on the stationary

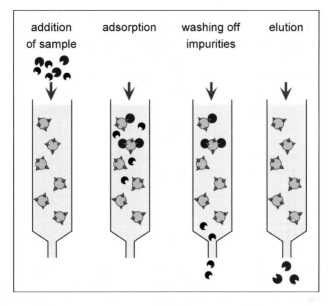

Fig. 2.9. Steps of affinity chromatography including sample addition, adsorption, washing and elution.

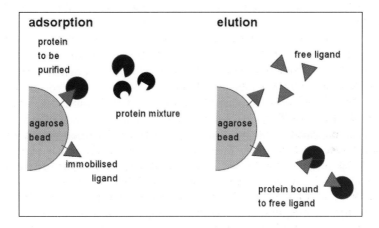

Fig. 2.10. Adsorption and specific desorption for affinity chromatography.

phase (Fig. 2.10, right). The free ligand competes with the bound ligand on the solid phase for binding sites on the protein. Once bound to the free ligand, the protein is eluted from the column. The separation matrix can then be regenerated for further use.

Table 2.4. Binding partners for affinity chromatography.

Immobilised ligand	Adsorbent
group-specific ligands	
protein A	IgG (binds via Fc region)
protein G	IgG
lectins	polysaccharides, glycoproteins
Cu^{2+}, Zn^{2+}	metall binding proteins
biotin	streptavidin
heparin	coagulation factors
nucleic acids	kinases, dehydrogenases, complementary sequence
monospecific ligands	
antibodies	antigens (viruses, cells)
enzyme inhibitors	enzymes
hormones	receptors

Ligands used for affinity chromatography can be divided into mono-specific and group-specific ligands (Table 2.4). *Monospecific ligands* show affinity for only one analyte. For example, a particular peptide hormone will only bind to its specific binding receptor. The peptide can be synthesised and immobilised onto the stationary phase and then used to isolate the receptor. Affinity chromatography is often the only method in which small quantities of a biomolecule can be isolated selectively. *Group-specific ligands* bind similar proteins that belong to the same protein class. For example, immobilised lectines can bind glycoproteins, glycolipids and polysaccharides. Another example is immobilised protein A, which can bind with the so-called Fc region of an antibody molecule (see section 5.1). The Fc region occurs in all antibody molecules. A wide range of group-specific ligands can be purchased commercially. Monospecific ligands often have to be synthesised in-house and then covalently bound to the matrix material of the stationary phase.

2.3.4   *Size Exclusion Chromatography (SEC)*

In size exclusion chromatography (SEC), dissolved molecules are separated according to their *size*, which is closely related to their *molecular weight*. This method can be applied for the separation of polymers in non-aqueous solutions, which is sometimes referred to as *gel permeation chromatography (GPC)*. It can also be used for the separation of biomolecules in aqueous solutions. Then the method is referred to *gel filtration chromatography*.

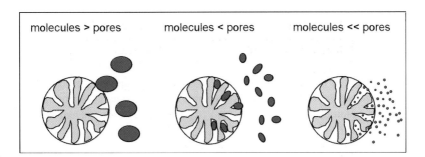

Fig. 2.11. The principle of size exclusion chromatography: Large molecules do not enter the pores and pass the matrix unretained, very small molecules spend a long time in the pores, but there is no differentiation between molecular sizes. Only within a critical size range is there a relation between residence time and molecular size, which can be used for separation.

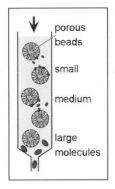

Fig. 2.12. Large molecules are un-retained and eluted first, smaller molecules are retarded by the pores of the stationary phase.

The chromatographic column is filled with a *porous* material such as a polymeric gel or agarose beads with diameters of typically 10 to 40 µm. Separation occurs, if the pore size is comparable to the size of the molecules passing through them (Figs. 2.11 and 2.12). Large molecules cannot enter the pores. They pass the matrix unretained and elute together with the solvent front. Smaller molecules enter the pores and have an average residence time, which depends on the size and shape of the molecule. The smaller the molecule, the longer its residence time in the pores and the greater its retention. Molecules that are much smaller than the pore size can enter the pores and have long residence times. However, there is no differentiation between molecular sizes anymore. Hence, all these small molecules are eluted together after a long retention time.

Differentiation and separation only occurs over a certain range of molecular sizes, typically between molecular weights of 2 kDa and 200 kDa, although this can be increased up to 1,000 kDa by the use of more specialised gels. This size range is dependent on the sizes of the pores and pore size distribution in the gel matrix. Retention volumes V_R are often used in size exclusion chromatography instead of retention times t_R. The *total volume* V_t of the separation column is the sum of the volume of the *gel particles* V_g, the volume of the solvent inside the pores, also called the *intrinsical volume* V_i and the volume of the free solvent outside the pores, the *inter particle volume* V_0:

$$V_t = V_g + V_i + V_0 \qquad \text{(equation 2.8)}$$

All analyte components are eluted *between* V_0 and $V_0 + V_i$. This can be understood by looking at the calibration curve in Fig. 2.13. The logarithm of the molecular weight, lg MW, is plotted against the retention volume V_R. Molecules larger than the *exclusion limit* are not retained, because they are too big to enter the pores. They are eluted together, producing one peak (peak A) at V_0. Molecules smaller than the limit of *total permeation* enter the pores completely and are strongly retarded. They are eluted at $V_0 + V_i$ producing a single band at the end of

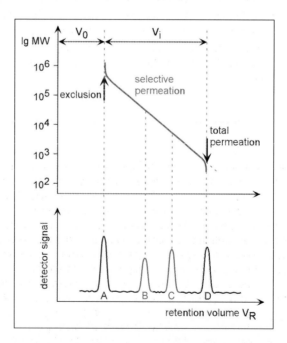

Fig. 2.13. Molecules are selectively retarded if their molecular weight is between the exclusion limit and the limit of total permeation.

the chromatogram (peak *D*). Molecules with molecular weights between these two values are distributed between the pores and the free solvent. Depending on their size, they spend more or less time in the pores. The selective permeation results in separation and hence individual peaks for each component (peaks *B* and *C*).

Unlike other chromatographic methods the mobile phase acts just as a solvent. Its physical properties do not influence the separation process. Solvent gradients do not alter the retention volume. As all compounds leave the column between V_0 and $V_0 + V_i$, no sample is lost on the stationary phase. The mobile phase usually consists of an aqueous buffer with an ionic strength of 50 to 100 mM. Typical flow rates are in the order of 0.1 to 1 mL min^{-1}.

Gel filtration chromatography is a very gentle method, as no harsh pH or ionic strength environments are required. One popular *application* is the separation of proteins from low molecular weight compounds such as peptides and amino acids. Another application is the separation of biomolecules such as proteins, fatty acids or nucleotides from each other according to their molecular size. Larger amounts of protein can be separated by gel filtration chromatography than can be separated by gel electrophoresis (section 3.2), although, the separation efficiency is not as high. Size exclusion chromatography can also be used to determine molecular weights. Calibration with samples of known molecular weight is necessary. Other methods such MALDI-TOF/MS (section 4.1) are very accurate and fast and often better suited for MW determination. However, mass spectrometers can only be considered as semi-quantitative detectors. SEC with UV detection, on the other hand, is a quantitative analysis method which allows for MW determination together with sample quantitation.

Summary

Chromatography is a separation method in which the analyte is contained in a mobile phase and pumped through a stationary phase. Sample components interact differently with these two phases and elute from the column at different retention times t_R. Chromatographic separations can be described quantitatively with a number of parameters including the capacity factor k', the selectivity factor α, the plate number N or height equivalent of a theoretical plate H and the resolution R_S. The optimum flow rate of a chromatographic separation can be determined with the van Deemter equation. In bioanalytical chemistry, chromatography is mainly employed for the isolation and purification of proteins. Reversed phase chromatography can separate biomolecules according to their interaction with the hydrophobic stationary phase and the hydrophilic moblile phase. This separation method can be coupled to an ESI mass spectrometer. Ion exchange chromatography separates molecules depending on their net charge. Affinity chromatography makes use of molecular recognition between biomolecules; and size exclusion chromatography allows for the separation of molecules depending on their size.

References

1. D. A. Skoog, F. J. Holler and T. A. Nieman, *Prinicples of Instrumental Analysis*, 5th edtion, Brooks Cole, 1997.
2. P. Bailon, G. K. Ehrlich, W.-J. Fung and W. Berthold (editors), *Affinity Chromatography: Methods and Protocols*, Humana Press, 2000.
3. D.T. Gjerde, C. P. Hanna and D. Hornby, *DNA Chromatography*, Wiley-VCH, 2001.
4. J.-C. Janson and L. Ryden (editors), *Protein Purification, Principles, High Resolution Methods and Applications*, 2nd edition, Wiley-VCH, 1998.

Chapter 3

ELECTROPHORESIS

In this chapter, you will learn about...

♦ ... the basic principles of separation in electrophoresis.

♦ ... the different concepts of gel and capillary electrophoresis.

♦ ... the different modes of gel and capillary electrophoresis that are employed in the separation of biomolecules.

Electrophoresis is a separation method based on the difference in mobility that analytes exhibit in an applied electric field. Electrophoretic methods can be applied to the separation of a wide variety of samples, including proteins, nucleic acids, amino acids and carbohydrates. Separations can be carried out on an analytical as well as preparative scale. The main advantage over liquid chromatography methods is the higher efficiency of electrophoretic separations. This is especially true for large molecules.

The first electrophoretic separation was published in the 1930s by the Swedish scientist Arne Tiselius. He separated serum proteins according to their charge in a u-shaped tube filled with buffer. For this and other accomplishments, Tiselius was awarded the Nobel prize in 1948. Today, electrophoresis is performed either on *slab gels* or in narrow bore *capillaries*. Slab gel electrophoresis was developed in the 1950s and has since become a standard technique used in biochemistry laboratories. Capillary electrophoresis (CE) was developed much more recently, in the 1980s. CE can easily be automated and is now an extremely fast-growing method in academia and industry alike.

Many different modes of electrophoresis have been developed to be performed both in gels and capillaries. The focus here is set on the techniques frequently applied to the analysis of DNA and proteins. This chapter is divided into three main sections. First, the basic principles and theory of electrophoretic separation are explained. The next section discusses gel electrophoretic instrumentation and techniques. These include sodium dodecyl sulfate – polyacrylamide gel electrophoresis (SDS–PAGE), isoelectric focussing (IEF) and 2D-gel electrophoresis (2D-GE). In the third section, the instrumentation and techniques for capillary

47

electrophoresis are described, including capillary zone electrophoresis (CZE), capillary isoelectric focussing (CIEF), micellar electrokinetic chromatography (MEKC) and capillary gel electrophoresis (CGE).

3.1   Principle and Theory of Electrophoresis

Electrophoresis is the movement of electrically charged particles or molecules in a conductive medium under the influence of an applied electric field. The conductive medium is usually an aqueous buffer, also referred to as an *electrolyte* or *run buffer*. The mixture of analytes is introduced into the medium containing the run buffer and an electric field is applied. In the example shown in Fig. 3.1, the analyte mixture contains negatively charged molecules. Upon application of the electric field, the anions start moving towards the positive electrode (anode). Differences in charge and size lead to different mobilities and thus separation of the different sample components. Similarly, positively charged ions move towards the cathode in an applied electric field.

Electrophoretic separations can be performed in free solution or in a solution containing a *non-conductive matrix* such as an *agarose* or *polyacrylamide gel*. For free solution based separations, narrow bore capillaries are used. The separation of the analyte ions occurs due to differences in mobility, i.e. differences in the charge to size ratio. Two analytes can only be separated if they have different charge to size ratios. Joule heating can interfere with the separation and cause band broadening. In a gel matrix this is prevented, as the gel acts as a heat dissipating medium and hence band broadening is minimised. The separation of analytes in a gel is also based on differences in mobility. Additionally, the gel has a *sieving effect*. Larger compounds are retarded more than smaller compounds. This means that in gel electrophoresis, two compounds with the same *charge to size ratio* can be separated as long as they are different in size.

The efficiency of an electrophoretic separation is governed by two main factors: the *electrophoretic mobility* (μ_{ep}) of the analytes and the so-called *electroosmotic*

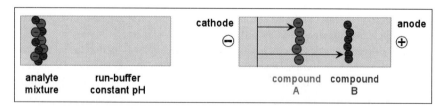

Fig. 3.1. The separation principle of electrophoresis. Particles with different charges, in this case negative charges, and different sizes migrate at different velocities in an applied electric field.

flow (EOF) of the bulk solution. Upon application of the electric field, a current passes through the conductive buffer, which leads to *Joule heating*. This effect must be limited or controlled as heating can interfere with the separation process. In the following sections, the principles of electrophoretic mobility, electroosmotic flow, Joule heating and separation efficiency are outlined in more detail.

3.1.1 *Electrophoretic Mobility*

The electrophoretic mobility, μ_{ep}, is a specific parameter for a given compound. It determines the velocity of a compound in an applied electric field. Compounds with different μ_{ep} can be separated from each other.

When an ion of charge q is placed into an electric field E, it experiences an electric force F_{ef}:

$$F_{ef} = q \cdot E \qquad \text{(equation 3.1)}$$

This electric force accelerates positively charged ions towards the negative electrode (cathode). Negatively charged ions are accelerated towards the positive electrode (anode).

The movement of the ions is opposed by the frictional force, F_{fr} of the medium molecules. This force is directly proportional to the radius of the ion, r, and its electrophoretic migration velocity, v_{ep}, as well as the viscosity of the medium, η.

$$F_{fr} = 6 \cdot \pi \cdot \eta \cdot r \cdot v_{ep} \qquad \text{(equation 3.2)}$$

In a constant electric field, equilibrium is reached between the frictional and the electric force (equation 3.3). Hence, ionic particles move with a constant velocity, v_{ep} (equation 3.4).

$$F_{ef} = F_{fr} \qquad \text{(equation 3.3)}$$

$$v_{ep} = \frac{q \cdot E}{6 \cdot \pi \cdot \eta \cdot r} \qquad \text{(equation 3.4)}$$

The electrophoretic mobility, μ_{ep}, is defined as the ratio of the migration velocity, v_{ep}, over the electric field strength, E (equation 3.5). μ_{ep} is more commonly used than v_{ep} to describe the migration of ions.

$$\mu_{ep} = \frac{v_{ep}}{E} = \frac{q}{6 \cdot \pi \cdot \eta \cdot r} \qquad \text{(equation 3.5)}$$

For a given separation medium the viscosity η is constant. Hence, the electrophoretic mobility, μ_{ep}, describes a charge to size (q/r) ratio. As mentioned earlier, two ionic species can be separated from each other on the basis of their q/r ratio, as they move at different velocities in an electric field. Changes in the buffer pH alter the charge of an analyte and thus its electrophoretic mobility.

3.1.2 *Joule Heating*

Upon application of the separation voltage, an electrical current passes through the conductive electrolyte buffer. This causes ohmic heating, also referred to as *Joule heating*. The heat must then dissipate through the walls and surfaces of the capillary or gel. A temperature gradient is formed across the capillary diameter or the gel cross section. This leads to *convective flows* within the electrolyte resulting in band broadening and loss of separation resolution. Moreover, high temperatures can cause thermal degradation of temperature sensitive analytes such as proteins.

To minimise Joule heating, two approaches can be used. (1) The generation of heat can be reduced by applying a lower electric field and by decreasing the conductivity of the separation buffer. However, lower separation voltages result in increased separation times and poorer separation efficiency. (2) A more commonly used approach is to improve the dissipation of heat by using small diameter capillaries or thin gels. These have a large surface-to-volume ratio, allowing heat to dissipate more quickly. Furthermore, the gel plates or capillaries can be enclosed in a thermostatically controlled environment so that a relatively stable temperature can be maintained. The electrical resistance of a narrow capillary or a thin gel is higher. According to Ohm's law, if the resistance is increased, the amount of electrical current flowing through the capillary or gel is reduced for a given voltage, which leads to less Joule heating. It must be noted that narrow capillaries or thin gels can only be loaded with a small volume of sample, which has a detrimental effect on the limit of detection.

3.1.3 *Electroosmotic Flow (EOF)*

Many of the materials used for electrophoretic separations such as glass, fused silica and agarose exhibit surface charges. For most bioanalytical separations, the pH of the buffer is basic so that negatively charged analyte ions are produced. The silanol (–Si–OH) groups on the surface of the capillary dissociate if the pH is > 4. Hence, the surface of the capillary becomes negatively charged (Fig. 3.2). This leads to a potential difference between the capillary surface and the bulk solution.

At the surface of the capillary, an *electric double layer* is formed. The negative surface charges are compensated by positive ions from the buffer solution. These

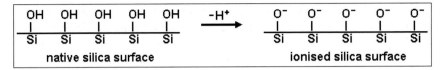

Fig. 3.2. The silanol groups on the surface of fused silica are deprotonated at pH > 4 and are thus negatively charged. At lower pH values, they are protonated.

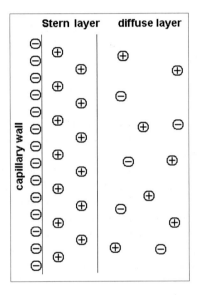

Fig. 3.3. The electric double layer consisting of the rigid Stern layer in proximity to the capillary surface and the diffuse layer extending into the bulk solution.

accumulate in close proximity to the negative surface charges and form a rigid layer, the *Stern layer* (Fig. 3.3). Within the Stern layer, there is a linear drop in potential. The Stern layer does not completely negate the surface charge of the capillary. Hence, a *diffuse layer* of mobile cations is formed next to the Stern layer. This diffuse layer extends into the bulk of the solution. The lower the ionic strength of the bulk solution, the thicker the diffuse layer. Typical values are in the order of nm to μm. The potential drop inside the diffuse layer is exponential. The *zeta-potential*, (ζ), can be approximated as the potential at the boundary between the Stern and the diffuse layer; ζ is commonly between 10 and 100 mV.

Upon application of an electric field, the cations in the diffuse layer move towards the cathode and drag the bulk solution with them. This movement of the bulk solution is called *electroosmotic flow*, (*EOF*). In gel electrophoresis, the phenomenon of EOF is also referred to as *electroendoosmosis*. The velocity of the EOF, v_{EOF}, is directly proportional to the dielectic constant ε and the zeta potential, ζ, of the buffer as well as the strength of the applied electric field, E. v_{EOF} is inversely proportional to the viscosity, η, of the separation buffer (equation 3.6):

$$v_{EOF} = \frac{\varepsilon \cdot \zeta \cdot E}{4 \cdot \pi \cdot \eta} \qquad \text{(equation 3.6)}$$

Similarly to equation 3.5, the *electroosmotic mobility*, μ_{EOF}, is defined as:

$$\mu_{EOF} = \frac{v_{EOF}}{E} \qquad \text{(equation 3.7)}$$

Often, the direction of the μ_{EOF} of the bulk solution is against the direction of the μ_{ep} of the analyte ions. This is because for most biomolecular separations, the analyte ions are negatively charged and will be dragged towards the anode, whereas the EOF is directed towards the cathode.

The *flow profile* of the EOF has the form of a *plug* (Fig. 3.4). The flow velocity is identical over the whole capillary diameter, except for the slower moving diffuse layer close to the capillary wall. This homogeneous velocity distribution minimises band broadening and, thus, increases separation efficiency. A radically different situation occurs with the pressure driven flow used in liquid chromatography. Here, the flow profile is parabolic; the flow velocities have a large distribution over the column diameter. Analytes in the middle flow considerably faster than analytes

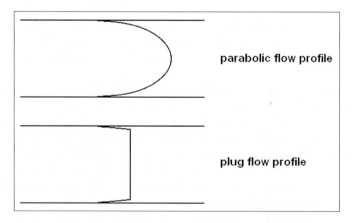

parabolic flow profile

plug flow profile

Fig. 3.4. A parabolic flow profile occurs in pressure driven flow such as in liquid chromatography, whereas the EOF flow profile has the form of a plug.

closer to the channel walls. This results in band broadening and loss of separation efficiency.

The velocity of the EOF is usually not very reproducible. Charges on the wall of the capillary can undergo electrostatic interaction with sample ions causing them to adsorb to the wall. Not only does this lead to sample loss, it also changes the velocity of the EOF. Separation efficiencies can thus become very irreproducible. In gel electrophoresis, the EOF is avoided by using gel media with no or very few surface charges. In CE, however, the EOF is quite often used as an essential parameter to optimise the separation process. If the EOF is predominant over the electrophoretic mobility, than all analytes, even negatively charged species, will be dragged towards the anode (section 3.1.4). This allows separation of positive, neutral and negative species. To achieve reproducible separations, it is essential to regulate the EOF.

Control of the EOF

In capillaries, the EOF can be controlled by (1) operating at low pH, (2) chemical surface modification, (3) dynamic coating of the capillary walls for example with a polymer layer or (4) by using additives that change the viscosity, η, and the zeta potential, ζ.

Surface charges are neutralised by operating at a pH low enough to protonate the silanol groups. This however only occurs at $pH < 4$, a pH at which many biomolecules are not stable.

Chemical modification of the silanol groups on the capillary walls can be performed to render them very hydrophilic or very hydrophobic. A hydrophilic surface, obtained by treatment with sulphonic acid, maintains a constant, high EOF. Hydrophobic functional groups attached to the surface lead to suppression of the EOF. A problem with chemical modifications is that their long-term stability is often very poor.

An alternative is *dynamic coating* of the channel walls by adding a polymer such as polyethylene glycol (PEG) to the run buffer. A polymeric viscous layer is formed on the capillary walls, which masks the charges and suppresses the EOF.

Neutral *additives* such as hydroxy ethyl cellulose or polyvinyl alcohol increase the viscosity of the run buffer and thus reduce the v_{EOF}. Furthermore, they suppress analyte-surface interactions. Similarly, organic solvents such as methanol and acetonitrile can be used to reduce or increase the viscosity respectively. Cationic surfactants such as dodecyl trimethyl ammonium bromide (DoTAB) adsorb onto the capillary walls and thus change the surface charge. This reverses the direction of the EOF. Surfactants must be used at low concentrations to avoid the formation of micelles, which may interfere with the separation process (section 3.3).

3.1.4 *Separation Efficiency and Resolution*

Efficiency and resolution of an electrophoretic separation are influenced by the electrophoretic flow as well as the EOF. The *apparent mobility*, μ_{app}, of an analyte is determined by the sum of its electrophoretic mobility, μ_{ep}, and the electroosmotic mobility, μ_{EOF}:

$$\mu_{app} = \mu_{ep} + \mu_{EOF} \qquad \text{(equation 3.8)}$$

If the EOF dominates over the electrophoretic motion, then all analyte components move towards the cathode. They are separated from each other due to their difference in apparent mobilities (Table 3.1, Fig. 3.5). Cations have the highest μ_{app}, because their μ_{ep} is in the same direction as the μ_{EOF}. Neutral species have no μ_{ep}. Hence, they are dragged along at the velocity of the bulk solution. Anions reach the cathode last, as their μ_{ep} opposes the direction of the μ_{EOF}.

Table 3.1. The sequence of elution at the cathode in case of predominating EOF: cations are eluted first, followed by neutral species and anions.

Charge	μ_{ep} and μ_{EOF}	μ_{app}	Sequence of elution
cation	same direction	$\mu_{app} = \mu_{ep} + \mu_{EOF}$	first
neutral	$\mu_{ep} = 0$	$\mu_{app} = \mu_{EOF}$	second
anion	opposite direction	$\mu_{app} = -\mu_{ep} + \mu_{EOF}$	last

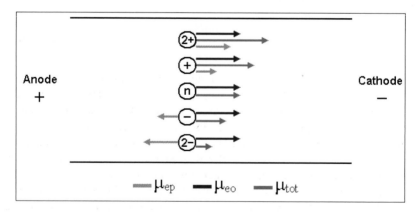

Fig. 3.5. Apparent mobilities of cations, neutral molecules and anions during electrophoresis with predominant EOF.

The *migration velocity*, v, of an analyte is defined as the product of its apparent mobility, μ_{app}, and the applied electric field strength, E, (equation 3.8). The field strength, E, is the ratio of applied voltage, V, over capillary length, L. Hence, v can be expressed as:

$$v = \mu_{app} \cdot E = (\mu_{ep} + \mu_{EOF}) \cdot \frac{V}{L} \qquad \text{(equation 3.9)}$$

The *migration time*, t, of an analyte, is defined as the capillary length, L, over the migration velocity, v. Substituting v for the expression in equation 3.9 leads to:

$$t = \frac{L}{v} = \frac{L^2}{\mu_{app} \cdot V} \qquad \text{(equation 3.10)}$$

Hence, the larger the applied field strength E, the faster the velocity and the shorter the migration time. This, however, is restricted by Joule heating (section 3.1.2).

In electrophoresis, band broadening is mainly caused by longitudinal diffusion. The *peak dispersion*, σ^2, is directly proportional to the diffusion coefficient, D, of the analyte and its migration time, t:

$$\sigma^2 = 2 \cdot D \cdot t = \frac{2 \cdot D \cdot L^2}{\mu_{app} \cdot V} \qquad \text{(equation 3.11)}$$

The *number of theoretical plates*, N, in electrophoresis can be approximated by:

$$N = \frac{L^2}{\sigma^2} = \frac{\mu_{app} \cdot V}{2 \cdot D} \qquad \text{(equation 3.12)}$$

The plate number is independent of the capillary length and the migration time. Large voltages lead to an increase in plate numbers. However, they also induce extensive Joule heating. Large molecules, with low diffusion coefficients, give high numbers of theoretical plates. For example, assuming $V = 25\,\text{kV}$ and $\mu_{app} = 2 \cdot 10^{-8}\,\text{m}^2\text{V}^{-1}\text{s}^{-1}$, the number of theoretical plates for a small cation such as potassium, K^+, is $N = 125{,}000$, whereas for a large protein such as albumin with a lower diffusion coefficient, $N = 4.2 \cdot 10^6$ is obtained.

In theory, millions of theoretical plates per metre can be achieved with electrophoresis making this technique superior to LC, where the number of plates per column is typically in the order of tens of thousands (section 2.2). In practice, however, lower plate numbers are observed in electrophoresis. Band broadening is caused by sample injection, Joule heating and adsorption of analytes to the separation matrix leading to plate numbers in the order of hundreds of thousands rather than millions.

The *resolution*, R_S, between two ions is dependent on the difference in electrophoretic mobility between the two species $\Delta\mu_{ep}$ (i.e. the separation selectivity), the applied separation voltage V, the apparent electrophoretic mobility, μ_{app}, and the diffusion coefficient D:

$$R_S = \Delta\mu_{ep} \cdot \sqrt{V} \cdot \sqrt{\frac{1}{\mu_{app}} \cdot \frac{1}{4 \cdot \sqrt{2 \cdot D}}} \qquad \text{(equation 3.13)}$$

The best way to optimise electrophoretic resolution, R_S, is to increase the separation selectivity, $\Delta\mu_{ep}$. This can be achieved for example by changing the pH of the run buffer or by changing the mode of electrophoresis (see later in this chapter). Increasing the separation voltage, V, also increases resolution. However, there is only a square root dependency with voltage and high voltages also lead to excessive Joule heating. Decreasing the apparent mobility, μ_{app}, results in an extended analysis time. The diffusion coefficient D is constant for a particular analyte in a given buffer. Resolution is higher for large molecules as these have small diffusion coefficients.

3.2   Gel Electrophoresis (GE)

In gel electrophoresis, separation takes place in an electrically non-conductive hydrogel medium such as agarose or polyacrylamide (PA), containing an electrolyte buffer. The pores of the gel function as a *molecular sieve,* which retards the migrating molecules according to their size. Furthermore, the gel acts as an *anti-convective* support medium, which minimises the diffusion of sample molecules and, thus, reduces band broadening. Hence, high plate numbers and high resolutions can be achieved, especially for high molecular weight molecules such as DNA or proteins. A very large number of compounds can thus be separated in a single run. As EOF is suppressed in gel electrophoresis, only analytes with a net charge can be separated. Neutral compounds do not migrate through the gel under the influence of an applied electric field. Gel electrophoresis is a rather slow and labour-intensive method, which is not readily automated.

A number of separation modes are possible. In *native gel electrophoresis*, charged analytes are distinguished according to their *apparent mobility*, μ_{app}, and size. In *sodium dodecylsulfate–polyacrylamide gel electrophoresis* (SDS–PAGE) analytes are treated such that they all exhibit the same charge to size ratio. Hence, they are separated only by differences in their size. This method can be used for molecular weight determination. In *isoelectric focussing*, IEF, amphoteric analytes such as peptides and proteins are separated according to their isoelectric point, pI. A combination of two separation modes in *two dimensional gel electrophoresis* (2D-GE) allows separation of thousands of proteins in one experiment.

These different modes are described in more detail in the following sections after an overview of gel electrophoresis instrumentation.

3.2.1 *Instrumentation for Gel Electrophoresis*

A power supply, an electrophoresis chamber with buffer reservoirs and a cooling thermostat are the major components required for gel electrophoresis (Fig. 3.6). The separation can be performed vertically or horizontally.

The *power supply* delivers voltages of typically 200 to 500 V with electrical currents of 400 μA to about 100 mA. The electrodes are dipped into the buffer reservoirs on each side of the gel. For most biomolecular separations, the pH is chosen such that the analytes are negatively charged. The analytes therefore migrate towards the anode. Hence, the electrode at which samples are introduced into the gel is chosen as the cathode, and the electrode at the other side of the gel is the anode. The whole instrument is encased in an insulating box to shield users from the high voltages.

The *electrophoresis chamber* contains the gel matrix immersed in an electrolyte buffer. Vertical gels can be polymerised in a glass or perspex tube with 0.5 to 1 cm i.d. and 3 to 10 cm length. Alternatively, the gel can be cast as a thin rectangular slab on which several samples can be run in parallel. The slab gel can be polymerised on an inert foil for horizontal separations or poured into a vertical tank. The thickness of slab gels is about 1–3 mm. A minigel has a length and width of about 8 cm × 8 cm, but larger gels of up to 40 × 20 cm are also commonly used.

For vertical separations, the samples are dissolved in a glycerol or sucrose solution of high density to prevent them from mixing with the buffer in the upper reservoir. The sample wells in the slab gel are made during casting by using appropriate combs or formers. Due to the shape of the sample wells, the analytes move in the form of wide narrow *bands*.

A *thermostat* is required for temperature control. Temperature controlled chambers ensure more reproducible separations as they help to dissipate heat from Joule heating (section 3.1.2) and protect sensitive analytes from thermal degradation.

Separation can take several hours. Once finished, the gel is removed from its holder. The analyte bands are then visualised, usually by staining. In the following paragraphs, gel media, sample preparation and band visualisation are discussed in more detail.

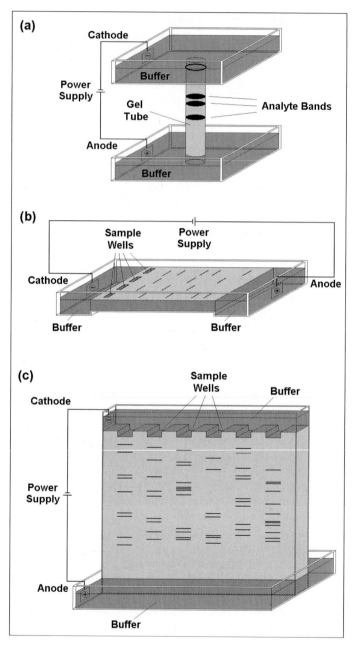

Fig. 3.6. Gel electrophoresis can be performed (a) in an upright tube. Alternatively, flat rectangular slab gels can be used which are positioned (b) horizontally or (c) vertically.

3.2.1.1   *Gel media*

Most commonly, agarose or polyacrylamide gels are used. These gels can be cast in the laboratory or obtained commercially. Many different formats are available from a variety of suppliers.

The gel pore size is an important parameter for electrophoretic separations. In *restrictive* gels, pores are small enough to act as *molecular sieves*. The electrophoretically migrating compounds are retarded according to their size, larger molecules move more slowly than smaller ones. In *non-restrictive* gels, the pores are too large to impede the sample movement. In this case, the migration time only depends on the mobility of the sample.

Agarose is a desulfonated derivative of agar, a compound, which can be isolated from algae cell membranes. For gel preparation, agarose is dissolved in the chosen buffer, heated to the boiling point and poured into the electrophoresis tank. Upon cooling, gelation occurs. The pores in agarose gels are fairly large, ranging from 150 nm at 1% concentration ($10 \, \text{mg mL}^{-1}$) to 500 nm at 0.16 % concentration ($1.6 \, \text{mg mL}^{-1}$). These large pores are only restrictive for very high molecular weight proteins or nucleic acids. For most analytes, agarose gels are *non-restrictive*. This can be useful for applications where molecular sieving is not desired, for example for isoelectric focussing (IEF) (section 3.4). Agarose gels are easy to prepare and they are non-toxic. However, these gels always have some surface charges, hence, they are never entirely free of *electroendoosmosis*. Additionally, they are not completely transparent, making sample readout more difficult.

Polyacrylamide (PA) gels are prepared by co-polymerisation of acrylamide and the cross-linking agent N,N'-methylene-bisacrylamide in the chosen electrophoresis buffer (Fig. 3.7). The pore size and hence the molecular sieving properties of PA gels depends on the *total gel concentration T%* (equation 3.14) as well as the *degree of cross-linking C%* (equation 3.15). T% values between 5 % and 20 % are commonly used. The higher the T% the more restrictive the gel. For example, at 5 % T and 3 % C the pore size is about 5 nm, whereas at 20 % T pore size decreases to about 3 nm for the same C%. Compared to agarose gels, pore sizes are much smaller. PA gels are generally *restrictive* and act as *molecular sieves*. High molecular weight compounds with MW > 800 kDa cannot be run on PA gels, as the pore size even at low T% values is not large enough.

$$\text{T\%} = \frac{Acrylamide(\text{g}) + Bisacrylamide(\text{g})}{100 \, \text{mL}} \cdot 100 \qquad \text{(equation 3.14)}$$

$$\text{C\%} = \frac{Bisacrylamide(\text{g})}{Bisacrylamide(\text{g}) + Acrylamide(\text{g})} \cdot 100 \qquad \text{(equation 3.15)}$$

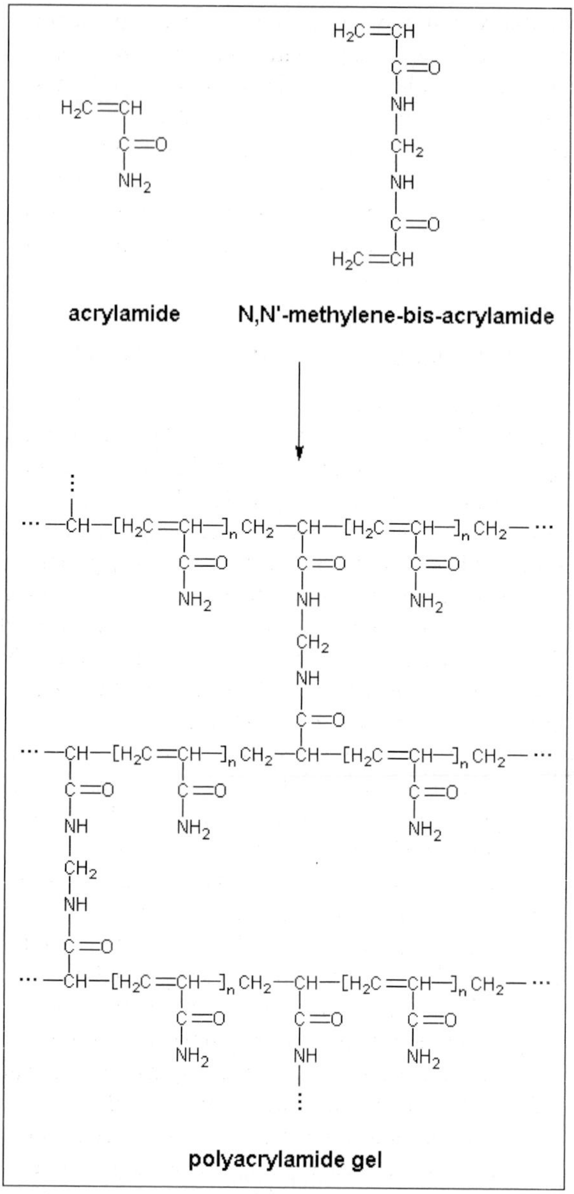

Fig. 3.7. Polyacrylamide gels are synthesised by co-polymerisation of acrylamide and N,N'-methylene-bisacrylamide.

It is also possible to cast polyacrylamide gels with a gradient in pore size. This gradient can either be linear or exponential. *Pore size gradients* can be achieved by continuously changing the monomer concentration whilst casting the gel.

Polyacrylamide gels are chemically inert and stable over a wide range of pH, temperature and ionic strength. The gels are transparent and a number of dye reactions can be used to visualise the separated bands. The gel surface exhibits hardly any charges, hence, electroendoosmosis is extremely low. Care must be taken during gel preparation. The monomers are both neurotoxins and potentially carcinogenic. Also, the free radical polymerisation is rather hazardous.

3.2.1.2   *Sample preparation and buffer systems*

The sample is dissolved in a solution of high density, usually containing glycerol and applied to the sample wells. To avoid blocking of the gel pores, samples should not contain any solid particles. High salt and buffer concentrations in the sample can interfere with the electrophoretic separation. Hence, these should be kept to a minimum, typically < 50 mM. The amount of sample required depends on the detection method, typical amounts are in the order of μg. The sample volume depends on the size of the sample well, which could range from μL to mL.

The buffer must be chosen such that the analyte molecules are charged, stable and soluble. To obtain negatively charged proteins, buffers with high pH values are used. Typical examples include Tris-glycine, pH 9.1 and Tris-borate, pH 8.3. Buffer concentrations are typically 50 to 100 mM.

A number of additives can be used to *increase solubility*. These include non-ionic surfactants such as Triton X-100 and zwitter ionic surfactants such as 3-[(cholamidopropyl) dimethyl ammonio]-1-propane sulfonate (CHAPS), which is particularly useful for solubilising membrane proteins.

Biopolymers can be separated in their native folded state or they can be denatured prior to separation. Native molecules with the same, sequence and length but different folding are likely to have different mobilities leading to broad bands or even different bands. To ensure reproducible and effective separations, a *denaturing agent* may be added to the buffer, the gel or both. This causes unfolding of the biopolymer, so that molecules with the same sequence and length will also have the same size and migrate as one band. Urea is an example of a denaturing agent. It is added to the buffer with concentrations as high as 8 M. Cationic and anionic surfactants at concentrations of about 10 % may also be used for denaturation. An example of a cationic surfactant is cetyl trimethyl ammonium bromide (CTAB). The most commonly used anionic surfactant is *sodium dodecyl sulfate (SDS)*, which is used for SDS–PAGE (section 3.2.3). To achieve complete unfolding, it is often required to heat the protein solution and to employ β-mercaptoethanol, an agent that cleaves sulfide bridges.

3.2.1.3 *Visualisation and detection*

The separated bands are most commonly visualised by staining with coloured, fluorescent dyes or with silver. A commonly used dye is *Coomassie brilliant blue*. The gel is immersed into an acidic alcoholic solution of this dye at elevated temperatures and left for a few hours. Excess dye is then removed by a number of washing steps. Limits of detection for proteins range from about 100 ng to 1 µg. *Silver staining* is more sensitive with detection limits < 1ng. The staining process is similar to photography. Silver nitrate, $AgNO_3$, is reduced to elementary silver, Ag^0, reduction occurs faster in proximity to the proteins in the gel. To obtain visible bands, the staining reaction must be stopped before all silver cations are reduced otherwise the gel would be stained completely. Very sensitive detection can also be achieved with radioactively or fluorescently labelled analytes as well as with enzyme labels that catalyse a reaction in which a coloured product is formed.

An example of a gel separation of DNA fragments from a soil bacterium is shown in Fig. 3.8. In this case, a fluorescent label was used to visualise the nucleic acids. On one gel, 18 samples and one standard mixture were run in parallel. The standard mixture contains molecules with known base pair lengths. The separation pattern obtained from the standard mixture is also referred to as a *ladder*. The number of base pairs in the sample DNA fragments can be estimated by comparing their bands to the bands in the ladder.

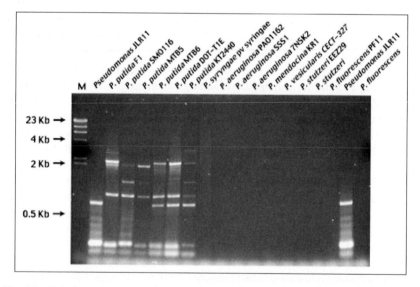

Fig. 3.8. Gel electrophoresis of PCR amplification products of genomic DNA of various strains of the soil bacterium *Pseudomonas putida*. Reprinted with permission from Nucleic Acids Res.2002, 30: 1826–1833, copyright 2002, Oxford University Press.

The stained gels can be scanned or photographed. Quantification is achieved by densitometry. With this method, the colour intensity of each band is determined by moving a light source over the gel and measuring the absorption.

Often recovery of the separated molecules is necessary for further analysis or sequencing (chapters 6 and 7). For this, a process called *electroblotting* is commonly employed. A polymeric membrane is placed over the gel and the molecules are transferred onto this membrane by an applied electric field. Sequencing reactions can be performed directly on the membrane. Furthermore, the membrane can be placed directly onto the sample tray of a MALDI-MS (section 4.1) for mass analysis.

3.2.2 *Modes of Gel Electrophoresis*

In electrophoresis, charged compounds move in the applied electric field towards the anode or cathode. In absence of band broadening factors, the compounds move in the form of *zones* or *bands*. Usually basic buffers are employed to obtain negatively charged analytes that migrate towards the anode. Charged analytes are separated according to differences in their electrophoretic mobilities, μ_{ep} (equation 3.5), i.e. differences in their charge to size ratios. EOF is suppressed in gel electrophoresis. Hence, the apparent mobility, μ_{app} (equation 3.8) is equal to the electrophoretic mobility μ_{ep}. In a restrictive gel with small pore diameters, the migration velocity also depends on the analyte size, which is closely related to the molecular weight. Thus, in the case of identical μ_{ep}, the molecular weight of proteins and nucleic acids can be estimated.

Pore size gradient gels are often employed for molecular weight determination. The gels can be produced such that the pore size decreases either linearly or exponentially (section 3.2.1.1). The analyte molecules migrate according to their electrophoretic mobility. However, they are eventually stopped by the ever decreasing pores. Large molecules are stopped after only a short migrating distance, whereas smaller molecules can migrate further, until they encounter a pore size too small for them to pass through. When all analyte molecules have reached their endpoint, the separation can be stopped and the molecular weight of the bands can be estimated by comparison to a standard mixture with known molecular weights.

3.2.3 *Sodium Dodecyl Sulfate–Polyacrylamide Gel Electrophoresis (SDS–PAGE)*

The separation principle of SDS–PAGE is solely based on the difference in protein size and hence molecular weight. The proteins are totally denatured in the presence of the anionic detergent sodium dodecyl sulfate (SDS). This detergent binds to proteins in a constant ratio of 1.4 g SDS per 1 g of protein; that is approximately one SDS molecule per two amino acid residues. The SDS-protein complexes assume a rod-like shape, with the large negative charge of SDS masking the intrinsic charge of the proteins, so that all SDS treated proteins have approximately a constant net charge per unit mass. Hence, all proteins have the same electrophoretic mobility. Separation in a polyacrylamide gel containing SDS, *SDS–PAGE*, depends only on the molecular sieving effect of the gel pores, i.e. on the radius of the analyte, which approximates its molecular mass. The larger the MW, the more slowly the protein migrates. Within a certain range, the relative mobilities of proteins are inversely proportional to the logarithm of the relative molecular mass.

Sample preparation usually involves heating the proteins to 95 °C, in the presence of excess SDS and a thiol-reducing agent such as β-mercapto ethanol. This results in complete unfolding of the tertiary and secondary structure. The molecules are stretched, sulfide bridges are cleaved and SDS binds to the amino acids.

SDS–PAGE is often used to determine the molecular weight of proteins. It is common practice to run a standard mixture with compounds of known molecular weight next to the samples for direct comparison of migration distance. This is often referred to as a molecular weight *ladder*. Typically molecular weights from about 15 to 200 kDa can be estimated with an accuracy of about 5 to 10 %.

3.2.4 *Isoelectric Focussing (IEF)*

Isoelectric focussing allows the separation of zwitterionic analytes such as proteins or peptides according to their *isoelectric point, pI* (section 1.1.1.1). IEF is applied to the separation and purification of proteins, peptides and amino acids on an analytical as well as preparative scale. The pI of a protein depends on the sum of all charges, as well as the 3D structure and post translational modifications such as phosphorylation, glycosilation and changes in oxidation state. The pI is, thus, a valuable parameter for studying post-translational modifications of proteins.

3.2.4.1 *The principle of IEF*

Instead of using a buffer with a constant pH over the whole gel, in IEF a *pH gradient* is generated in which the pH value increases smoothly from anode to cathode. When a zwitterionic compound such as a protein is placed into this pH gradient it migrates until it reaches a point where its net charge equals zero. This is the position in the gradient, where the pH equals the protein's pI (Fig. 3.9). If the pH is larger than the pI, then the protein has a negative net charge and it moves towards the anode. If the pH is smaller than the pI, the protein becomes positively charged and moves towards the cathode. If the pH equals the pI, the protein is

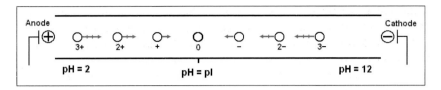

Fig. 3.9. The principle of IEF: When a protein is placed in a pH gradient and a voltage is applied, it migrates towards its isoelectric point.

not charged and, hence, it does not move in the electrical field anymore. If the molecule diffuses into any direction, it becomes charged again and moves back to its pI. The sample can be introduced anywhere in the pH gradient and it will always move to exactly the same point.

IEF is an endpoint method. Once the band is focussed, it will not be affected by any band broadening over time. Bands as narrow as 0.01 pH units can be obtained.

The *resolution* of isoelectric focussing can be expressed as ΔpI, the minimum pH difference to resolve two compounds. If compounds have similar pIs, resolution can be improved by using pH gradients with a narrower range and smaller increments. Other factors that influence ΔpI include the applied electric field strength E, the diffusion coefficient D of the protein and the change of its mobility at different pH.

3.2.4.2 Instrumental considerations for IEF

Agarose or PA gels are both used for IEF. Sufficiently large pores are required as this reduces the molecular sieving effect and hence shortens the separation time. With agarose gels proteins larger than 800 kDa can be focussed. Polyacrylamide gels are limited to lower MW.

Proteins often precipitate during isoelectric focussing because (1) they are concentrated into a sharp band and (2) they do not exhibit any net charge at their pI and can thus undergo extensive hydrophobic interaction. To improve solubility, a number of additives can be used. The denaturing agent urea may be used at concentrations of up to 6 M. Non-ionic and zwitterionic surfactants are also commonly used.

3.2.4.3 Formation of pH gradients

For good separation results, a stable pH gradient with constant conductivity is required. This can be achieved by carrier ampholytes or immobilised pH gradients.

Carrier ampholytes are synthesised low molecular weight oligomers with a number of basic amino groups and acidic carboxyl groups. A mixture of hundreds of different carrier ampholytes, each with a different pI is used to form a pH gradient. The ampholytes must have a high buffering capacity at their pI. It is also important that each single compound in the ampholyte mixture has the same concentration and that there are no gaps in the spectrum of pIs.

The gel is immersed into a buffer with a medium pH which contains 1 to 2 % of a carrier ampholyte mixture. The anode reservoir is filled with a low pH buffer, called the *anolyte*. The pH of the anolyte has to be more acidic than the lowest pI in the ampholyte mixture. The cathode reservoir is filled with a high pH buffer, the *catholyte*. The pH of the catholyte has to be more basic than the highest pI in the ampholyte mixture. At the beginning, the pH in the gel equals the medium pH of the buffer and almost all ampholyte molecules are charged. The basic carrier ampholytes are negatively charged and the acidic ampholyte compounds are positively charged. When an electric field is applied, the basic ampholytes move towards the anode. Similarly, the positively charged ampholytes move towards the cathode. This means that the cathodic site of the gel becomes more acidic and the anodic site more basic. The ampholyte in the mixture with the lowest pI moves towards the end of gel on the anodic site. When it migrates further, it is exposed to the low pH of the anolyte. It becomes positively charged and thus moves back into the gel. Similarly, the ampholyte with the highest pI moves towards the end of gel on the cathodic site. If it moves any further, it becomes negatively charged from the catholyte and is thus forced to move back into the gel towards the anode. The other carrier ampholytes arrange themselves between these two in the sequence of their pIs. This results in the formation of a continuous pH gradient (Fig. 3.10).

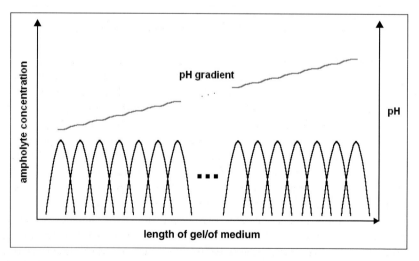

Fig. 3.10. A pH gradient with increasing pH over the gel length formed by a mixture of hundreds of ampholytes each with a different pI. The concentration of each ampholyte is the same to ensure a homogeneous conductivity.

Due to the high buffering capacity of the ampholytes, the pH gradient is stable even in the presence of larger concentrations of analytes. However, sometimes the gradient starts to drift over time, often towards the cathode. This compromises the performance of IEF.

The choice of carrier ampholytes determines the pH range of the gradient. Often hundreds or even thousands of different compounds are used. The pH gradient may be between 2.5 and 11, however, narrow range pH gradients give better resolutions.

The protein sample can be introduced at any place in the gradient. It is usually introduced into the gel together with the ampholyte solution. The protein is then focussed at the same time as the pH gradient develops. Due to the low diffusion coefficients of proteins, they form much sharper bands than the ampholytes themselves. The ampholyte molecules have relatively low molecular weights (< 1 kDa) and thus high diffusion coefficients. They form relatively broad bands that overlap (Fig. 3.10).

A pH drift can be avoided by using *immobilised pH gradients (IPG)*. These are produced by incorporate substances called immobilines in the gel polymerisation process. Immobilines are not zwitterionic; they are either acidic or basic. The pH depends on the ratio of these immobilines. Controlled mixing during gel casting is, thus, required to obtain a good pH gradient.

For both carrier ampholytes and IPGs, the gradient can be linear or non-linear, depending on the anticipated separation. The pH gradient can be measured by running a mixture of markers with known pI in parallel to the analyte mixtures as a reference.

3.2.5 *Two-Dimensional Gel Electrophoresis (2D-GE)*

In 2D-GE, two electrophoresis modes are combined on a single gel. One separation is performed in the first dimension, followed by another separation perpendicular to the first one. With this method, mixtures with thousands of proteins or nucleic acids can be separated with high resolution. The resulting "fingerprint" can be compared to electronic databases. Results can, however, be difficult to reproduce and this technically challenging method requires experienced operating personnel.

The parameters that govern the separation process such as charge, mobility, size and pI depend on the particular electrophoresis mode used. It is not uncommon that after separation one band contains more than one species. To resolve these analytes, a second separation can be performed, which separates the analytes in terms of a different parameter than in the first separation. For example, a mixture of proteins can be separated in the first dimension according to their isoelectric point by IEF. The first separation is performed in a single lane. In the second dimension, the proteins are then separated by SDS–PAGE according to their molecular weight. Assuming that each separation is capable of resolving 100 bands, a 2D method can resolve, in theory, up to 10,000 zones.

2D-GE can be performed on small plates with an area of a few cm^2, but larger plates, which measure 45 × 30 cm are also commonly used. After separation, the analytes must be visualised, for example by staining. The gels are then scanned and can be compared to electronic databases. 2D-GE is commonly applied to proteins, DNA and RNA.

3.2.5.1 *Separation of proteins*

Complex mixtures of proteins and peptides can be separated by 2D-GE. Protein samples can be taken directly from a cell culture. A single protein can be also partly digested to analyse its fragments. Typically 1,000 – 2,000 proteins can be resolved in a single run, in some cases, the separation of up to 4,000 proteins is possible. Usually IEF is employed in the first dimension followed by SDS-PAGE in the second dimension. In Figure 3.11 an example is shown of a 2D-GE separation of a protein mixture form the soil bacterium *Burkholderia cepacia*. The separation in the first dimension (left to right) was performed with an immobilised pH-gradient

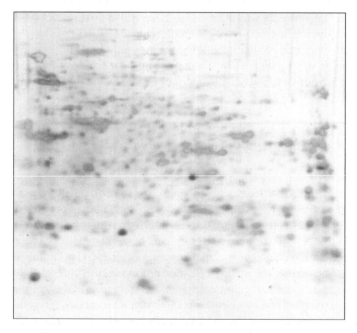

Fig. 3.11. 2D-GE of proteins from the soil bacterium *Burkholderia cepacia*. The separation in the first dimension (left to right) was achieved by IEF. In the second dimension, the proteins were separated according to their molecular weight in a polyacrylamide gel with a pore size gradient (Courtesy of B. Crossett, Centre for Molecular Microbiology and Infection, Imperial College London, UK.)

(IPG) from pH 3 to pH 10. The separation in the second dimension was performed on a polyacrylamide gel with a pore size gradient from 4 to 12 T%, which differentiates the proteins according to their molecular weight in the range from $\sim$150 kDa to $\sim$6 kDa.

3.2.5.2 *Separation of nucleic acids*

Long DNA fragments such as whole genes are usually analysed after partial digestion with a restriction enyzme (section 6.3.1). Often two such digestion steps are necessary to completely resolve all DNA fragments into individual bands. The first separation is carried out after treatment with one restriction enzyme. The bands obtained are then treated with another restriction enzyme and separated in the second dimension. Transfer between the two dimensions is the most challenging step in the separation process. The bands can be treated *in situ* with the second restriction enzyme, either by soaking them in a solution containing the enzyme or by incorporating the restriction enzyme into the gel. The DNA fragments can also be transferred onto a cellulose membrane by blotting, then treated with the restriction enzyme, and finally transferred onto the second dimension gel.

2D electrophoresis of RNA usually employs a gel in native conditions in one dimension, i.e. the RNA molecules retain their native folding configuration. In the second dimension the gel contains a denaturing agent such as SDS. Thus, the RNA molecules undergo a change in conformation and can be separated from each other.

3.3 Capillary Electrophoresis (CE)

Capillary electrophoresis is based on the same principle as gel electrophoresis. Charged analytes can be separated in an applied electric field according to their mobility. In contrast to gel electrophoresis, however, separations are carried out in a small diameter capillary containing a free solution of electrolyte rather than on a slab gel. Moreover, convective flows due to Joule heating occur more easily in a free solution than in the gel. In contrast to GE, electroosmotic flow is often part of the separation process.

Capillary electrophoresis has a wide applicability. High molecular weight compounds such as proteins, nucleic acids and oligosaccharides can be separated as well as smaller biomolecules such as peptides and amino acids. CE is not restricted to charged analytes. Neutral molecules can be separated from each other by employing a variation of CE called micellar electrokinetic chromatography (MEKC). This is frequently used for the separation of chiral drugs in pharmaceutical research.

Capillary electrophoresis was developed in the 1980s by James Jorgenson and Krynn Lukas. They separated derivatised amino acids in a 75 μm inner diameter

Table 3.2. A number of modes of capillary electrophoresis with their commonly used abbreviations, separation principle and applications.

Separation method		Separation principle	Applicable to
Capillary Zone Electrophoresis	CZE	charge and size	ions
Capillary Isoelectric Focussing	CIEF	isoelectric point, pI	zwitter-ions
Micellar Electrokinetic Chromatography	MEKC	charge and lipophilicity	ions, ion pairs, neutral species
Capillary Gel Electrophoresis	CGE	size (MW)	ions

capillary with high efficiency. Today, CE is a fast growing method finding increasing use in academia and industry due to a number of advantages that CE has over neighbouring techniques such as HPLC (chapter 2) and gel electrophoresis (chapter 3.2). The sample volumes required are very low, in the order of nL. Solvent consumption is also very low, in the order of a few mL per day. The separation times in CE are very short in comparison to both LC and GE. Resolution and efficiency of capillary electrophoretic separations are very high due to the plug flow profile (section 3.1.3) and the high voltages that can be applied; up to 30 kV are commonly used. These factors have lead to the name *high performance capillary electrophoresis (HPCE)*. In addition to these advantages, CE can be completely automated including sample injection, separation and data analysis.

Similarly to gel electrophoresis, a number of modes can be employed that separate analyte mixtures according to different properties. Some of these modes are summarised in Table 3.2 together with their commonly used abbreviations. Depending on the principle of separation, different species can be analysed. After an overview of capillary electrophoretic instrumentation, the different modes of capillary electrophoresis are described in more detail including *capillary zone electrophoresis (CZE), capillary isoelectric focussing (CIEF), micellar electrokinetic chromatography (MEKC)* and *capillary gel electrophoresis (CGE)*.

3.3.1   *Capillary Electrophoresis Instrumentation*

The instrumentation required for Capillary Electrophoresis is relatively simple. A typical example of a CE apparatus is shown in Fig. 3.12, consisting of vials

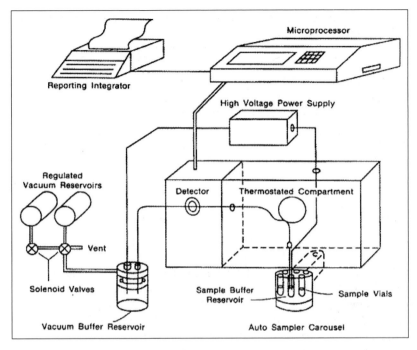

Fig. 3.12. Schematic of a typical CE instrument, courtesy of Applied Biosystems.

with samples and buffer, a high voltage power supply, a capillary enclosed in a thermostatically controlled compartment, an on-column detector and a data output system as well as a vacuum system for sample injection.

The capillary is filled with buffer solution and dipped into buffer vials at both the source and destination end. For sample injection, the source end of the capillary is temporarily placed into the sample vial. For separation, the capillary is dipped back into the buffer vial. A high voltage is applied and the sample components start to migrate towards the electrode at the destination end. Migration occurs with different velocities according to the electrophoretic mobilities of the analytes as well as the velocity of the EOF. The capillary has a detection window towards its outlet. The sample components pass this detection window and their signals are plotted as peaks in an *electropherogram*, with the migration time on the x-axis and the signal intensity on the y-axis, similar to a chromatogram (section 2.2).

The *power supply* generates a high voltage over the capillary by means of two platinum electrodes, which are dipped into the buffer reservoirs at each end of the capillary. The applied electric fields in CE can be much higher than in gel electrophoresis. The reason for this is that less Joule heating is produced in the narrow capillaries and a better heat dissipation is achieved due to the large surface to volume ratio. Typically, voltages up to 30 kV are applied, with electrical currents up to about 300 μA. For a 100 cm long capillary this results in an electric field of $E = 300\,\mathrm{Vcm}^{-1}$. Depending on the type of sample, the analytes are either positively or negatively charged. In order to avoid repositioning of the detector, the power supplies must be capable of reversing the polarity. Safety interlocks must be fitted into all instruments to prevent the operator from coming into contact with the high voltages.

The *capillaries* used in CE have internal diameters of 20 to 100 μm and outer diameters of about 400 μm. They are typically between 10 and 100 cm long. The most popular material is *fused silica*, i.e. amorphous quartz, which is transparent to UV and visible light. These capillaries are externally coated with a polyimide layer of about 10 μm thickness to increase flexibility. For the detection window, this polyimide layer must be removed, by either scratching or burning it off.

The capillary is usually enclosed in a thermostatically controlled environment for temperature control. This is because the viscosity of the buffer varies with temperature and Joule heating must be dissipated effectively to avoid temperature fluctuations, which can have dramatic effects on the efficiency and reproducibility of CE separations.

The source and destination vials as well as the inside of the capillary are filled with a *buffer*, also referred to as carrier electrolyte or background electrolyte. The purpose of the buffer is to maintain the pH as well as the conductivity during the electrophoretic separation. A controlled pH is crucial for maintaining a constant net charge on the biomolecules and, thus, maintaining their electrophoretic mobility μ_{ep}. A controlled conductivity is required so that Joule heating can be controlled. Buffer concentrations in CE are typically in the order 10–100 mM.

The *injection* system must be capable of reproducibly introducing very small sample volumes into the capillary. The volume of the whole capillary is only in the order of μL. To minimise band broadening, sample plugs must be as short as possible. Hence, not more than a few nL of sample are introduced into the capillary. Two injection methods are commonly used: (1) electrokinetic injection and (2) hydrodynamic injection.

In *electrokinetic injection*, voltages are used to introduce the sample into the capillary. The source end of the capillary together with the source end electrode are placed into the sample solution. A high voltage is applied over the capillary between the sample vial and the destination vial for a given period of time. This causes the sample to move into the capillary according to its apparent mobility, μ_{app} (equation 3.8). After sample introduction, the capillary and electrode are returned into the buffer reservoir and CE separation is started. A problem associated with electrokinetic injections is that during the injection process, discrimination occurs between different components in the sample. Components with higher μ_{app} are injected in larger quantities than those with lower μ_{app}. One way to minimise this discrimination is to use lower voltages for injection than for the later separation.

Hydrodynamic injection can be performed in three different ways. (1) In pressure injection, a precisely controlled external pressure is used to force a controlled amount of sample into the capillary. (2) In vacuum injection, a vacuum is applied to the buffer reservoir at the detector end of the capillary for a controlled period of time at a regulated reduced pressure. (3) For gravity flow injection, the sample vial with one end of the capillary is elevated to a certain height above the other end of the capillary for a given period of time. Gravity forces a sample plug into the capillary.

Hydrodynamic injection can be advantageous over electrokinetic injection as there is no inherent discrimination of the injected sample components. Electrokinetic injection is generally easier to incorporate into a CE instrument. The reproducibility of injection is often not very high, as the injection volume depends on the ionic strength of sample buffer and the sample.

To shorten sample plugs and to increase sensitivity, a method called *sample stacking* can be employed. One example of how to achieve this is by dissolving the sample in a buffer with a much lower conductivity than the running buffer. A sample plug is hydrodynamically injected into the capillary and the separation voltage is applied. Due to the low conductivity of the sample buffer, the field strength across the sample plug is considerably higher than in the running buffer. This causes the ions to migrate faster, until they reach the boundary with the running buffer. Upon entering the running buffer, they migrate more slowly under the influence of the weaker field. This process continues until all the ions in sample zone reach the boundary, resulting in the concentration or focussing into a smaller zone.

The main challenge for CE *detectors* is the small diameter of the capillary and the small sample volumes encountered. Detection schemes employed for capillary electrophoresis include measurement of UV absorption, fluorescence and refractive index. Electrochemical signals and conductivity as well as radioactivity from radioisotopes have also been measured. The signals obtained are plotted against the migration time in the form of an *electropherogram.* In recent years, coupling of CE to a mass spectrometer (CE-MS) has been achieved.

UV-absorption detection at a chosen wavelength is most commonly used. Peptides are usually measured at $\lambda = 210$ nm, proteins and DNA at $\lambda = 260$ nm or $\lambda = 280$ nm (see section 1.3.3). The absorbance is measured directly through a detection window in the capillary approximately 1 mm long. For this, the polyimide coating of the capillary has to be removed. The small capillary diameter, less than $100 \, \mu$m, results in a short detection path length and thus in low sensitivity. This problem may be partly overcome by use of a *Z-cell* (Fig. 3.13), which effectively increases optical path length 10 to 15 times. However, due to band broadening within the cell, the separation efficiency is decreased.

Diode array detectors (DAD) can take whole UV/vis spectra of the sample whilst it is passing through the detection window. Several spectra per second can

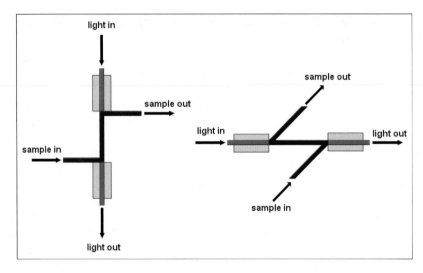

Fig. 3.13. Different types of Z-cells used for UV-absorbance detection. The sensitivity is improved due to the increased optical path length.

be recorded. The information content of this detection method is much higher than just recording the absorbance at one wavelength. However, the sensitivity is much lower.

Very high sensitivity can be obtained by measuring *laser induced fluorescence (LIF)*. In most analyte mixtures, not all the components are naturally fluorescent. Thus, derivatisation of the analytes with a fluorescent marker is necessary for detection. It is essential, that all sample components are homogeneously derivatised. Sometimes derivatives are not very stable, so care has to be taken.

Conductivity, electrochemical, radioisotopic and refractive index detection are more rarely used.

Similarly to LC (section 4.3.4), also CE instruments can be coupled to a *mass spectrometer*, providing a powerful system for analysis of complex samples. The output of the electrophoresis capillary is connected to an electrospray ionisation (ESI) source, *CE-ESI-MS*. Usually a make up flow is necessary to increase the flow rate for a stable spray. To avoid contamination of the ion source, it is essential to utilise only volatile buffer components such as ammonium acetate and volatile additives such as methanol, acetonitrile and acetic acid.

Limits of detection for CE are generally between mg L^{-1} (ppm) for a UV-vis detector and μg L^{-1} (ppb) for a fluorescence detector. These are roughly 100 to 1000 times lower than in liquid chromatography. *Qualitative analysis* of sample components via CE is usually based on the migration time of the sample in the capillary and possibly on the spectra obtained. In almost all cases, it is necessary to run a standard with known samples for comparison. *Quantitative analysis* can be performed according to peak height or peak area after calibration with a series of samples of known concentration.

3.3.2 *Capillary Zone Electrophoresis (CZE)*

Capillary Zone Electrophoresis (CZE) is a widely used technique that is capable of separating anions as well as cations in the same run. The capillary as well as the buffer reservoirs are filled with the same electrolyte. A voltage is applied and the analyte ions move independent of each other with a different velocity according to their apparent mobility, μ_{app}. In contrast to gel electrophoresis, where EOF is generally minimised or avoided, a controlled electroosmotic flow is often used during a CZE separation.

If the *EOF* is *dominant* over the electrophoretic mobilities, then it is possible to separate cations, neutral species and anions in a single run (Fig. 3.14). The strong EOF causes the bulk of the solution to migrate towards the cathode. All analyte components are dragged with the bulk flow. The cations reach the detector first. They migrate with the sum of their electrophoretic mobiliy and the EOF. Cations with higher μ_{ep} reach the detector before cations with low μ_{ep}. Neutral components do not exhibit any electrophoretic mobility μ_{ep}, they move solely due to the EOF. Hence, they reach the detector after the cations. All neutral compounds have exactly the same mobility $\mu_{app} = \mu_{EOF}$. This means, that all neutral compounds elute in a single peak, without any differentiation. Negative sample compounds have an electrophoretic mobility μ_{ep} towards the anode. However, due to the dominant EOF, they are dragged towards the cathode as well and reach the detection window last. Anions with a low μ_{ep} reach the detection window before anions with a high μ_{ep}.

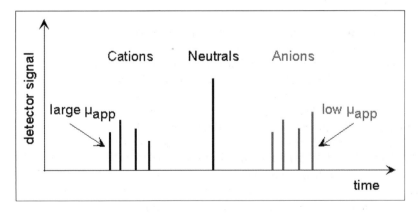

Fig. 3.14. Elution order in CZE with predominant EOF.

Capillary electrophoresis separations can be *optimised* by changing a number of parameters including pH and ionic strength of the buffer, additives, temperature and capillary coatings. A change in pH has an effect on the net charge and mobility of most biomolecules. The ζ-potential and, thus, the velocity of the EOF also depend on the pH. The ionic strength of the buffer influences the analyte mobilities, Joule heating and viscosity. Temperature control is essential for good heat dissipation to avoid convection within the capillary and to protect thermally instable analytes. The temperature also affects the mobility, dissociation constant and solubility of the sample. Coating of the capillary walls can be used to control or alter the EOF. The charged silanol groups on the capillary surface can be chemically modified or dynamically coated (section 3.1.3).

3.3.3   *Capillary Isoelectric Focussing (CIEF)*

Isoelectric focussing as outlined in section 3.2.4 for gels can also be performed in a capillary containing a free solution of electrolyte. *Capillary isoelectric focussing, (CIEF)* can be advantageous over IEF. The process is readily automated. Due to higher electric fields, the focussing times are very short; often the separation is finished within minutes rather than hours. Only minute sample quantities are required, a few µL in the sample reservoir and a few nL in the capillary. Quantification is also easier and more reproducible in CIEF.

Instrumentation for CIEF is not much different from other CE methods. The capillary is filled with an ampholyte mixture. An electric field of about 400 to $600\,\mathrm{Vcm^{-1}}$ is applied and the pH gradient develops inside the capillary. The sample can be placed together with the ampholyte mixture into the capillary and focussed whilst the pH gradient is developing. It is essential to coat the capillaries to prevent EOF as this may sweep the analytes past the detector before focussing has finished.

Detection in CE occurs at a fixed point and not over the whole medium as in gel electrophoresis. Hence, it is necessary to *mobilise* the *focussed bands* and move them past the detection window. This can be achieved by chemical and hydro-dynamic flow mobilisation in coated capillaries or by electroosmotic mobilisation in uncoated or partially coated capillaries.

For *electroosmotic mobilisation* uncoated or partially coated capillaries are used for example by adding 0.1 % methyl cellulose to the buffer. The EOF is suppressed but not completely eliminated. The bulk solution moves due to the EOF at the same time as focussing takes place. It is necessary to make sure that focussing occurs before the analytes are eluted past the detector. The advantage of this method is that the buffer does not need to be changed for mobilisation, nor does the voltage have to be turned off. However, uncoated capillaries can give rise to uneven pH gradients.

In *hydrodynamic mobilisation*, focussing and mobilisation are carried out as two separate steps. After focussing is completed, the capillary is connected to a pressure pump or vacuum while the focussing voltage is still applied. The focussed bands are thus, moved and pass the detector. The pressure driven flow results in a parabolic flow profile, which can compromise the resolution due to band broadening. An advantage of this method is that very basic and very acidic proteins, which are focused at the extremities of the capillaries, can be mobilised easily.

Another approach is *electrophoretic (salt) mobilisation*. Again the focussing and mobilisation steps are separated. After focussing is completed, the buffer composition at the anode or cathode is changed by the addition of a salt leading to a disrupted pH equilibrium. In case of *cathodic mobilisation*, a salt such as NaCl is added to the cathode buffer reservoir. The Cl^- ions compete with the OH^- ions to move into the capillary towards the anode. This results in a drop of pH. The proteins formerly at their pI, now experience a pH lower than their pI and become positively charged. This results in a movement of the proteins towards the cathode past the detection window. Similarly, in *anodic mobilisation*, a salt such as NaCl is added to the anode. The sodium cations compete with the H_3O^+ ions to enter the capillary and move towards the anode. As now less H_3O^+ ions enter the capillary, an increase in pH occurs. The proteins are now experiencing a pH that is higher than their pI and become negatively charged. They are, thus, dragged towards the anode. Electrophoretic mobilisation can be achieved with a high resolution as there are no compromising effects from discontinuous EOF or parabolic hydrodynamic flow.

Alternatively, imaging along the whole length of the capillary can eliminate the need for mobilisation.

3.3.4 *Micellar Electrokinetic Chromatography (MEKC)*

Separation in Micellar Electrokinetic Chromatography (MEKC) is based on partitioning of the analyte molecules between the aqueous run buffer and the core of micelles, which are contained in the run buffer. The technique is essentially a hybrid between CE and liquid chromatography (LC). The run buffer and micelles are moved through the capillary by an applied electric field. The analytes are dragged with the bulk solution. Similar to LC, the analytes partition between two phases, in this case two mobile phases, the hydrophilic run buffer and the hydrophobic micelles. Unlike other electrophoresis modes, MEKC can distinguish between different neutral compounds according to their hydrophobicity.

MEKC was developed in the 1980s by a Japanese scientist, Shigeru Terabe. The method was initially developed for the separation of neutral compounds, but it has proven to be capable of separating both neutral and ionic compounds. Application examples include separation of amino acids, oligopeptides, nucleic acids, fatty acids, steroids and pharmaceutical drugs.

3.3.4.1 *Principle of MEKC*

MEKC *instrumentation* is not different from the apparatus used for capillary zone electrophoresis (chapter 3.3.2). The only deviation is that the run buffer contains *micelles*. MEKC is sometimes also referred to as *micellar electrokinetic capillary chromatography* (MECC). The signals are recorded as an *electrokinetic chromatogram* with signal intensity versus time.

Micelles are made up from surfactant molecules. If the surfactant concentration in the aqueous run buffer exceeds a certain level, then the molecules arrange themselves in the form of micelles. This concentration is called the *critical micelle concentration (CMC)*. The most commonly used surfactant is the anionic detergent, SDS, $[CH_3\text{-}(CH_2)_{11}\text{-}O\text{-}SO_3]^- Na^+$. It contains a hydrophilic sulphate group and a hydrophobic $CH_3\text{-}(CH_2)_{11}$-group. When SDS is dissolved in an aqueous buffer, the molecules aggregate in the form of spherical micelles. The hydrophobic chains are orientated towards the centre of the aggregate, while the hydrophilic groups point outwards (Fig. 3.15). The spherical micelles thus have hydrophobic core and a hydrophilic surface. Not all SDS molecules in the solution are arranged in

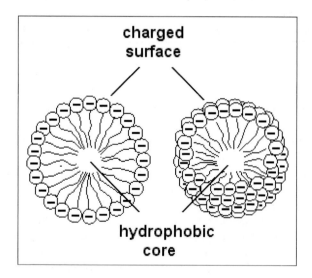

Fig. 3.15. Above the critical micelle concentration, CMC, surfactants form spherical aggregates with their hydrophobic tails coordinated towards the centre while the polar groups point outwards.

Table 3.3. Some common surfactants for MEKC and their CMC.

Detergent	CMC (mM)
Anionic	
sodium dodecyl sulfate (SDS)	8
sodium tetradecyl sulfate (STS)	2.2
sodium cholate (SC)	13–15
sodium taurocholate (STC)	10–15
Cationic	
cetyltrimethylammonium chloride (CTAC)	1.3
cetyltrimethylammonium bromide (CTAB)	0.92
dodecyltrimethylammonium bromide (DoTAB)	14–16
Zwitter-ionic	
3-[3-(cholamidopropyl) dimethylammonio]-1-propanesulfonate (CHAPS)	6–8
Non-ionic	
Triton X-100	0.2–0.9

micelles. In fact, there is a dynamic equilibrium between molecules being dissolved in the buffer solution and being organised in micelles.

A suitable *detergent* for MEKC must have a good solubility in the buffer, low UV absorption, a low viscosity and not too high a CMC to avoid extensive Joule heating. The detergent can be *anionic, cationic* or *zwitterionic*. Even *non-ionic* detergents can be employed as additives to form co-micelles. Some commonly used detergents are listed in Table 3.3 together with their critical micelle concentrations. The *aggregation number*, AN, is defined as the number of molecules that the micelle consists of. Sodium cholate micelles only consist of 2 to 4 molecules. With AN = 62, SDS micelles consist of quite a high number of molecules. Anionic surfactants typically have an alkyl chain of 8 to 14 carbon atoms in length. Surfactants with shorter alkyl chains have a high CMC, surfactants with longer chains exhibit solubility problems in aqueous solutions. Cationic surfactants, even at low concentrations, adsorb strongly to the negatively charged capillary walls. The charge of the capillary surface is effectively reversed from negative to positive and consequently the direction of the EOF is reversed. In this case, separations are run from the cathode towards the anode.

When an electric field is applied over a capillary containing an aqueous buffer with SDS micelles, then the EOF is directed towards the cathode. The surface of the SDS micelles is negatively charged. Thus, the micelles have an electrophoretic mobility towards the anode. Usually the EOF is dominant, hence, the micelles

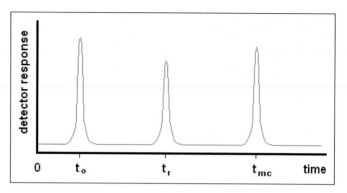

Fig. 3.16. In MEKC, neutral analytes pass the detection window at their retention time t_R, which is between the zero retention time of the bulk flow t_0 and the retention time of the micelles t_{mc}.

are also dragged towards the cathode, but their movement is retarded due to the electrophoretic mobility in the opposite direction.

The separation of neutral analytes is based on their *partitioning* between the aqueous run buffer and the hydrophobic micelle core. If a hydrophilic compound is injected into the capillary, it does not partition into the hydrophobic micelle core. The compound migrates with the bulk of the solution and passes the detection window at t_0 (Fig. 3.16). If a relatively hydrophobic compound is injected, it is completely incorporated in the hydrophobic core of the micelles, i.e. it is completely solubilised by the micelles. This relatively hydrophobic compound migrates with the velocity of the micelles and reaches the detection window at t_{mc}. A compound with intermediate hydrophobicity partitions between the aqueous buffer and the hydrophobic micelle core and passes, the detector at a particular retention time t_R which depends on the compound's hydrophobicity. The more hydrophobic the compound, the more time it spends inside the micelles, the longer is its retention time t_R. In case of neutral analytes, all sample components are detected between t_0 and t_{mc}. The ratio of t_{mc} to t_0 is defined as *elution ratio*, and the time between t_0 and t_{mc} is called the *migration window*. In the case of charged analyte molecules, the situation is different. Cations might move faster than the bulk solution and pass the detection window at a time shorter than t_0. Anions might be retarded more than the micelles and reach the detector at a time larger than t_{mc}.

It is possible to measure t_0 by adding a hydrophilic compound such as methanol to the buffer. Methanol does not partition into the hydrophobic micelles core. It is eluted at t_0. Totally hydrophobic compounds can be used to measure t_{mc}. For example, the dye Sudan III is completely solubilised by the micelles and passes the detector at t_{mc}.

3.3.4.2 Basic theory of MEKC

Similar to liquid chromatography, the capacity factor k', the selectivity α and the resolution R_S of an MEKC separation can be defined.

The *capacity factor k'* describes the partitioning of the analyte between the two phases. It is defined as the ratio of the time that the analyte spends in the aqueous phase over the time it spends in the micelle phase:

$$k' = \frac{t_R - t_0}{t_0 \cdot (1 - (t_R/t_{mc}))} \qquad \text{(equation 3.16)}$$

As the migration time of the micellar phase, t_{mc}, approaches infinity, i.e. the micellar phase becomes stationary, the expression for k' resumes the conventional chromatography form (equation 2. 1).

The *selectivity factor α* describes the relative velocities of the analytes with respect to the aqueous buffer. For MEKC, α is defined in exactly the same manner as in LC (see equation 2.2):

$$\alpha = \frac{k'_B}{k'_A} = \frac{t_{r,B} - t_0}{t_{r,A} - t_0} \qquad \text{(equation 3.17)}$$

The *resolution, R_S, of* two neighbouring peaks in an MEKC separation can be calculated as a product of an efficiency term, a selectivity term and a retention term:

$$R_S = \left(\frac{\sqrt{N}}{4}\right) \cdot \left(\frac{\alpha - 1}{\alpha}\right) \cdot \left(\frac{k'_B}{1 + k'_B} \cdot \frac{1 - (t_0/t_{mc})}{1 + (t_0/t_{mc}) \cdot k'_A}\right) \qquad \text{(equation 3.18)}$$

Increasing the number of theoretical plates, N, has only a limited effect on the resolution. A large migration window, i.e. a small ratio of t_0/t_{mc}, results in an improved resolution. The most pronounced effect on R_S can be achieved by increasing the selectivity α. In the following section, a number of possibilities on how to achieve an improved resolution are discussed.

3.3.4.3 Parameters influencing MEKC separations

One way to optimise a separation is by changing the composition of the aqueous mobile phase. For example, the *pH* of the buffer can be varied. This can have dramatic effects on the EOF and therefore change the zero retention time t_0. The charge of the micelles might also be altered by a change in pH, resulting in a different t_{mc}. Furthermore, the degree of the ionisation of the analytes and the electrolyte system depends on the pH. This influences both the capacity factor k' and the selectivity factor α.

Organic modifiers such as acetonitrile, dimethylformamide and tetrahydrofuran can also be added to the buffer system. They alter the viscosity of the buffer and thus the velocity of the electroosmotic flow (equation 3.6). *Non-ionic surfactants* such as Triton X-100 can combine with the ionic surfactants to form co-micelles. These co-micelles have different mobilities and thus a different t_{mc} than the original micelles.

The micelle generating *surfactant* can be changed. This is similar to changing the stationary phase in liquid chromatography. However, this can be done relatively easily and at much lower cost than in LC. The *micelle concentration* has an influence on analyte retention times. The CMC for SDS is 8 mM, and typical concentrations used for MEKC are between 25 and 150 mM. A high concentration of SDS molecules results in a large number of micelles being generated. The probability of an analyte partitioning into a micelle is thus much larger.

3.3.5   *Capillary Gel Electrophoresis (CGE)*

In capillary gel electrophoresis (CGE), the capillary is filled with a gel rather than with a free solution. The separation in CGE works on the same principle as slab gel electrophoresis: The analytes migrate depending on their electrophoretic mobility but they are retarded by the gel pores depending on their size. The gel acts as an anti-convective medium, reducing band broadening and as a molecular sieving material.

With CZE as described in section 3.3.2, it is often impossible to separate different nucleic acids from each other because they have a similar charge to size ratio and, thus, similar electrophoretic mobilities (equation 3.5). The same is true for SDS denatured proteins. Introducing a gel into the capillary, leads to an additional molecular sieving effect. Large analytes are retained more than smaller ones, enabling separation of analytes with similar mobilities.

The advantages of performing gel electrophoresis inside a capillary are numerous. The applied electric field strength over a capillary can be much higher than over a gel. This leads to *faster* separations (equation 3.10) and better separation *efficiency* (equation 3.12). CGE can be fully *automated* and *sensitivity* can be improved due to on-column detection. This is in sharp contrast to the labour intensive slab gel electrophoresis, which requires gel casting, staining, densiometry and scanning steps to be performed manually. On the downside, CGE does not enable parallel processing of several samples. The method can only be performed in one dimension, the powerful 2D-separations possible with gels (section 3.2.4) cannot be transferred to the capillary format. Furthermore, CGE works with very low sample amounts. It must be classified as an analytical method. Preparative separations are not feasible.

Applications of CGE include separation of ssDNA, dsDNA, RNA and proteins. Sizing of DNA fragments is a major application. This is important for DNA sequencing (section 6.3) as well as after PCR product analysis (section 6.2). SDS–PAGE of proteins (section 3.2.3) is also commonly performed in capillaries.

The gel media employed for CGE can be grouped into cross-linked and linear gels. *Cross-linked gels* such as cross-linked polyacrylamide (section 3.2.1.1) are very rigid. They are polymerised inside the capillary and covalently bound to the capillary wall. These gels are also referred to as chemical gels. Once polymerised, they cannot be removed from the capillary. Their lifetime is limited to about 100 separations. However, they exhibit a very good separation efficiency. *Linear gels* include methyl cellulose derivatives, polyethylene glycol (PEG), dextrose, agarose and linear polyacrylamide. They consist of linear polymer chains, which are held together via physical interactions. Hence, these gels are also referred to as physical gels or polymer networks. The viscosity of polymer networks is very high. They can be injected into the capillary under high pressure. The capillary walls are usually treated to suppress EOF, which might extrude the gel from the capillary.

Detection methods in gel filled capillaries are most commonly UV-absorption and laser induced fluorescence. Proteins have a high UV-absorbance at $\lambda = 200$ nm and $\lambda = 214$ nm. Polyacrylamide gels absorb strongly at $\lambda < 230$ nm and are, therefore, not suitable for protein analysis. Dextrane or PEG can be used as alternatives, as they are both UV-transparent. The absorption of DNA fragments is usually measured at $\lambda = 260$ nm (see section 1.3.3.1), for which PA gels are well suited.

Summary

Electrophoresis is the movement of charged species in an applied electric field. Factors influencing this movement are the electrophoretic mobility μ_{ep} of the analyte, the electroosmotic flow (EOF) of the bulk solution and Joule heating. These parameters are used to separate sample compounds from each other.

Electrophoretic separations can be carried out in a free solution or a solution containing an anti-convective gel matrix. EOF is usually negligible in the gel matrix, whereas in capillaries it is often dominant over the electrophoretic mobilities of the analytes. Gel electrophoretic separations can be used on a preparative scale. The separated DNA fragments or proteins can be blotted onto a membrane and used for further analysis such as sequencing. Capillary electrophoresis must be classified as an analytical method. It has advantages over gel electrophoresis, as it is readily automated and separations can be achieved in much shorter times due to higher voltages.

Both gel and capillary electrophoresis can be performed in a variety of modes, which separate the analytes due to different factors. Zone electrophoresis can be performed on gel and in capillary. The sample compounds are separated according to their size and mobility on the gel, whereas in the capillary they are only separated according to their mobility. In SDS–PAGE, proteins are separated only due to differences in their size. This method can also be performed on a gel as well as inside a capillary. Isoelectric focussing (IEF) is applicable for separating zwitterionic analytes with different pI. A pH gradient is formed inside the capillary or on the gel and the analytes move until they reach the point where the pH equals their pI. Two dimensional electrophoresis can only be performed on gels. It is the most powerful separation method as thousands of compounds can be resolved in a single run. MEKC can only be performed in capillaries. It is the only electrophoretic method that is capable of separating neutral analytes from each other according to their hydrophobicity.

Capillary electrophoresis and HPLC are often complementary methods. Both are capable of separating polar as well as non-polar samples, both can be used for high molecular weight compounds and both can be used in a wide range of pH. HPLC is more widely used, as it was developed before CE and many separation problems have been solved using HPLC. Method development is often considered as simpler for CE. However, as analytical chemists have only fairly recently been trained in CE, the method is still less popular than HPLC.

CE has a number of advantages over HPLC (section 3.3). Sample volumes are much smaller, solvent consumption is reduced, separation efficiencies are higher, and separation times are shorter. The main reasons for these are the plug flow profile observed in CE, in contrast to the parabolic flow profile that develops in HPLC and the high separation voltages that can be applied. Detection sensitivities are generally higher for HPLC, due to the larger diameter of the HPLC columns in comparison to the CE capillaries.

References

1. D. R. Baker, *Capillary Electrophoresis*, John Wiley and Sons, New York, 1995.
2. R. Weinberger, *Practical Capillary Electrophoresis*, 2nd edition, Academic Press, 2000.
3. S. J. Y. Li, *Capillary Electrophoresis, Principles, Practice and Applications*, Elsevier Science, 1993.
4. B. D. Hames (editor), *Gel Electrophoresis of Proteins, A Practical Approach*, 3rd edition, Oxford University Press, 1998.
5. D. Rickwood and B. D. Hames (editors), *Gel Electrophoresis of Nucleic Acids, A Practical Approach*, 2nd edition, IRL Press of Oxford University Press, 1990.

Chapter 4

MASS SPECTROMETRY

In this chapter, you will learn about...

♦ ...the basic principle and instrumentation of mass spectrometry (MS).
♦ ...the two most important techniques for the analysis of biomolecules: MALDI-TOF/MS and ESI-MS.
♦ ...how mass spectrometry is used to determine the molecular weight of even large biomolecules like DNA and proteins.
♦ ...how mass spectrometry is used as a separation method.
♦ ...and how mass spectrometry can be used to obtain structural information about peptides and polynucleotides.

Mass spectrometry is among the most powerful tools in protein and DNA analysis. It can determine molecular weights of biomolecules as large as 500,000 Da with high accuracy. Structural information like the amino acid sequence in a peptide or the sugar sequence in an oligosaccharide can be obtained. Some mass spectrometers can be coupled directly to a separation method such as LC or CE to combine the strengths of both techniques.

4.1 The Principle of Mass Spectrometry

A mass spectrometer determines the molecular weight of ions in vacuum. The sample molecules are first ionised in what is known as the *ion source* (Fig. 4.1). The gaseous ions are then introduced into a *mass analyser* and separated according to their *mass-to-charge (m/z) ratio*. A *detector* registers the signals and passes information to a computer for analysis and spectrum recording. To avoid collisions between ions and air molecules, a high vacuum of about 10^{-5} Pa is required.

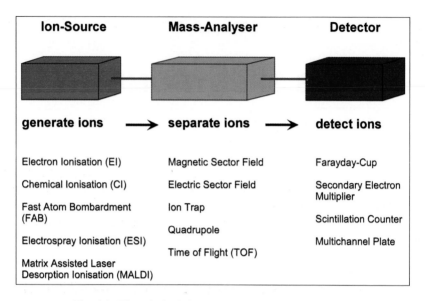

Fig. 4.1. The principal components of a mass spectrometer.

4.1.1 *Ionisation*

The sample molecules can be ionised by one of several techniques. In *electron impact ionisation, (EI)*, electrons are fired at the sample molecules, whereas in *chemical ionisation, (CI)*, the sample molecules are collided with a reactive gas. The sample can also be bombarded with argon atoms (*fast atom bombardment, FAB*) or the dissolved sample can be sprayed into an electric field (*electrospray ionisation, ESI*). Furthermore, the sample can be co-crystallised with a matrix and then ions can be generated by exposure to photons (*matrix assisted laser desorption ionisation, MALDI*).

All these techniques result in positively and/or negatively charged ions in the gaseous phase. *Hard ionisation methods* like EI and FAB lead to the breakdown of the sample molecules into smaller fragments. These fragments give a "fingerprint" of the sample and thus valuable information. *Soft ionisation methods* like ESI and MALDI lead to molecular ions $[M]^+$, and quasi molecular ions such as $[M + H]^+$ which can be used for molecular weight determination.

4.1.2 *Mass Analyser*

The mass analyser separates the ionised species according to their mass-to-charge (m/z) ratio. This can be achieved by magnetic or electric sector fields, an ion trap, a quadrupolar magnetic field with high frequency, or in a time-of-flight (TOF) analyser.

4.1.3 *Detector*

A Faraday-cup, a secondary electron multiplier, a scintillation counter or a multichannel plate are used for ion detection.

In bioanalytical chemistry, soft ionisation methods such as ESI and MALDI are preferred as they allow analysis of whole protein or DNA molecules. MALDI is usually combined with TOF analysers, whereas ESI is combined with quadrupole analysers. These two methods are explained in detail in the following sections.

4.2 Matrix Assisted Laser Desorption Ionisation – Time Of Flight Mass Spectrometry (MALDI-TOF/MS)

Koichi Tanaka presented experiments for *soft laser desorption ionisation (SLD)* of proteins in 1987. However, the predominant and most widely used version of SLD, *matrix assisted laser desorption ionisation, MALDI*, was introduced shortly afterwards by Michael Karas and Franz Hillenkamp. Tanaka was awarded the Nobel Prize for his cornerstone invention in 2002. Prior to that, no method was available to transfer large biomolecules with molecular weights of more than 1,000 Da into the vacuum without fragmenting them.

With MALDI-TOF, molecular weights above 500,000 Da can be determined with sensitivities as low as fmol and mass accuracies as high as 0.1–0.01 %. Furthermore, small amounts of contaminants are tolerated, sample preparation is fairly straightforward and the information obtained can be submitted automatically for a database search.

4.2.1 *Ionisation Principle*

The ionisation principle is based on the soft desorption of the solid sample molecules into the vacuum and subsequent ionisation. First, the sample is co-crystallised with a 1,000–10,000 excess of a suitable *matrix* on a metallic plate. Small, organic, UV-absorbing molecules like sinapinic acid are used as matrix materials (Table 4.1). An electric field is applied between the sample plate and the entrance to the time-of-flight analyser (Fig. 4.2). A pulsed laser beam is then

Table 4.1. Typical matrix substances used for MALDI in biochemical analysis.

Compound		Wavelength	Used for
2,5-Dihydroxy benzoic acid (DHBA)	COOH, OH, HO (structure)	337 nm 355 nm	peptides, proteins, oligosaccharides
Sinapinic acid (SA)	COOH, HO, OH, OCH$_3$ (structure)	337 nm 355 nm	proteins, peptides, glycoproteins
Nicotinic acid	COOH, N (structure)	266 nm	proteins, peptides, oligonucleotides
α-Cyano-4-hydroxy-cinnamic acid (α-CHCA)	NC—C=CH—COOH, OH (structure)	337 nm 355 nm	peptides, proteins, oligosaccharides
4-Hydroxy-picolinic acid	OH, N, COOH (structure)	337 nm 355 nm	oligonucleotides
Succinic acid	H$_2$C—CH$_2$, HOOC COOH	2.94 μm 10.6 μm	peptides, proteins
Glycerine	H$_2$C—CH—CH$_2$, OH OH OH	2.79 μm 2.95 μm 10.6 μm	peptides, proteins

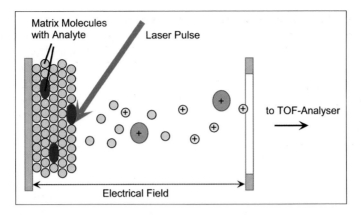

Fig. 4.2. The principle of matrix assisted laser desorption.

focussed onto the crystal. The matrix is chosen such that it absorbs readily at the laser wavelength. The sample, however, should not absorb at this wavelength. When bombarded with the photons from the laser pulse, the matrix molecules are excited rapidly and transferred into the gas phase together with the analyte molecules before energy is transferred to neighbouring molecules. Some matrix and analyte molecules become ionised during this process. Once in the gaseous phase, the ions are accelerated towards the TOF analyser by the applied electrical field. The matrix molecules can also take and donate protons or electrons to the analyte molecules and transfer ionisation energy. The analyte ions obtained are predominantly molecular ions $[M]^+$ or quasi-molecular ions like $[M + H]^+$ as well as adducts with alkali metal ions from buffer solutions like $[M + Na]^+$. Multiply-charged molecular ions also occur. Due to the high matrix concentration the analyte ions are prevented from interacting with each other. MALDI is a very soft ionisation method; large biomolecules like proteins, nucleic acids, polysaccharides and lipids can stay intact. Without the matrix, the analytes could only be desorbed at higher energies, but this would result in their fragmentation.

Nitrogen lasers with a wavelength of $\lambda = 337$ nm are most commonly used for matrices that absorb in the UV-area. Pulses are several ns long with photon energies of 3.68 eV. As an alternative, Nd-YAG lasers at $\lambda = 266$ or 355 nm are also used. IR-lasers are somewhat softer than UV-lasers but there is a limited choice of IR-absorbing matrices. Most frequently used are Er-YAG lasers at $\lambda = 2.94 \, \mu m$ with pulses less than 100 ns and photon energies of 0.42 eV or CO_2 lasers with $\lambda = 10.6 \, \mu m$. The laser beam passes through optical and electrical components and has a diameter of about $150 \, \mu m$ when it hits the target.

4.2.2   *Mass Analysis in Time-of-Flight Analyser*

The mass analyser used for MALDI is usually a time of flight (TOF) analyser, which allows for high resolution and accurate mass determination even for high molecular weight species. The ions desorbed by the laser pulse are accelerated in an electric field to a kinetic energy of several keV (Fig. 4.3). They then enter a field free tube in which they drift along with different speeds according to their mass/charge ratios. At the end of the tube, the ions hit a detector and the drift time is measured electronically to a high accuracy.

The kinetic energy of the drifting ions is defined as:

$$E_{kin} = \tfrac{1}{2} \cdot m \cdot v^2 = z \cdot e \cdot V \qquad \text{(equation 4.1)}$$

where m is the mass of the ion, v the velocity of the ion after the acceleration region, z is the ion charge, e the elementary charge and V the voltage of the applied electrical field. Light ions are accelerated more than heavier ions and reach the detector first. The velocity, v, can also be defined as the length of the field free drift tube, L, over the time of flight, t:

$$v = \frac{L}{t} \qquad \text{(equation 4.2)}$$

Substituting the velocity v in equation 4.1 by equation 4.2 leads to:

$$\frac{m}{z} = \frac{2 \cdot e \cdot V}{L^2} \cdot t^2 \qquad \text{(equation 4.3)}$$

The m/z ratio of the ion is proportional to the square of the drift time. Hence, the mass of an ion can be determined by measuring its drift time once the analyser is

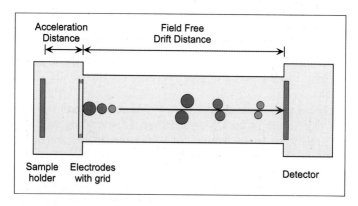

Fig. 4.3. The principle of a linear TOF analyser: the accelerated ions have different velocities according to their m/z.

calibrated with substances of known weight and charge. With this time measurement even heavy ions can be detected accurately, making TOF ideal for molecular weight analysis of biomolecules. Typically, the flight tubes have a length of about two meters, resulting in flight times in the order of microseconds. Very good sensitivities can be obtained, in comparison to other analysers, because all ions that have passed the pin hole into the time of flight analyser reach the detector.

Because not all the molecules get desorbed at the same time in exactly the same place, slightly different velocities are obtained for identical ions which can result in a broad peak and poor resolution. Using a reflector TOF set-up (Fig. 4.4) instead of the linear set-up shown above (Fig. 4.3), can overcome this problem.

In a *reflector TOF* tube, an electrical opposing field is applied at the end of the drift tube at which the ions are forced to change direction. Ions of the same mass but higher kinetic energies (velocities) get deeper into this opposing field and need more time for change of direction but catch up with smaller ions at a certain point in the drift field. When positioning the detector at this focusing point, very sharp signals can be obtained. Another advantage of the reflectron TOF is the capability of detecting ions that decay whilst in the tube, a process which is called *post source decay (PSD)*.

The linear tube allows detection of high molecular weights up to hundreds of kDa, but the resolution decreases as the mass increases. The reflector tube is limited to masses up to tens of kDa; resolution, however, is improved.

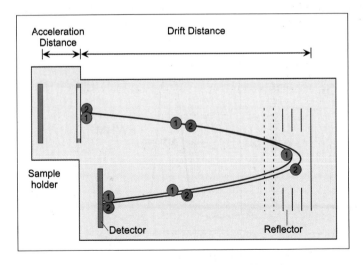

Fig. 4.4. The principle of a reflectron TOF analyser.

4.2.3   Detection of Ions

For ion detection in time of flight analysers, usually secondary electron multipliers are used.

4.2.4   Resolution

An important parameter for the quality of a mass analyser is its capability to separate ions with small mass differences Δm. This is described by the resolution, R_S, the ratio of the mass, m, over the difference Δm of an ion with mass $m + \Delta m$:

$$R_S = \frac{m}{\Delta m} = \frac{m_1}{m_2 - m_1} \qquad \text{(equation 4.4)}$$

Generally, the higher the resolution the better is the separation. But when are two peaks considered as being separated? This is a question of definition and depends on the analyser. For TOF, Δm is defined as the *full width at half maximum (FWHM)*, i.e. the width of the peak at half its height (Fig. 4.5). With this definition, it is possible to read R_S out of a single peak. Typical resolutions obtained for TOF instruments are $R_S = 15,000$ (FWHM). For other mass analysers, other definitions like the *10 % valley* or *50 % valley* are used (Fig. 4.6). For the 50 % valley definition, two peaks are considered separated if the minimum between them (the valley) is not more than 50 % of the peak height whereas for the 10% valley, the minimum between two peaks must not be more than 10% of the peak height.

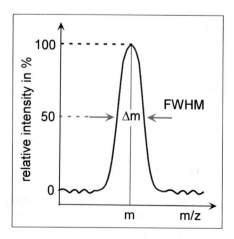

Fig. 4.5. Definition of resolution by full width at half maximum (FWHM).

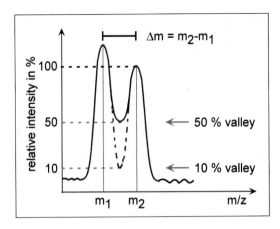

Fig. 4.6. Definition of resolution by 10 % and 50 % valley.

4.2.5 Sample Pretreatment

As mentioned earlier, to achieve MALDI, the sample molecules have to be co-crystallised with a high molar excess of an appropriate matrix. The *function of the matrix* is to absorb and accumulate the energy of the laser radiation and, thus, protect the analytes from destruction and fragmentation. A good matrix material must adsorb strongly at the laser wavelength. Additionally, the matrix must also be chemically inert, stable in vacuum and be able to embed the analyte. Furthermore, the matrix material should promote co-desorption of the analyte upon laser irradiation as well as ionisation of the analyte by donating protons. A number of compounds fulfil these requirements (see Table 4.1), however, their performance varies depending on the analyte, and some trial and error is required to find the optimum matrix for a specific analyte.

Sample preparation for MALDI is relatively straightforward. The sample could be a commercially obtained protein or peptide, or a band from a dried SDS gel (section 3.2.3) or a spot from a dried 2D gel (section 3.2.5). Solutions of the sample and the matrix are made up and mixed either in a tube prior to placing onto the target plate or on the target plate itself. To obtain good spectra, it is essential to keep the salt concentrations in buffers to a minimum. Two common methods are described below.

The *dried droplet method* is the method originally introduced by Hillenkamp and Karas (Fig. 4.7). A saturated matrix solution, 5–10 g L^{-1}, depending on the solubility of the matrix, is prepared in water, water-acetonitrile, or water-alcohol mixtures. In a second vessel, the sample is diluted to about 100 mg L^{-1} in a solvent that is miscible with the matrix solution. The matrix and sample solutions are then mixed such that the final molar ratio is 10,000 : 1 with a final volume of a few µL.

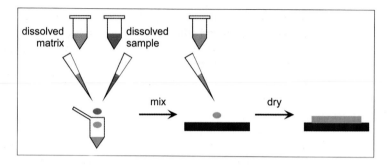

Fig. 4.7. The principle of the dried droplet method.

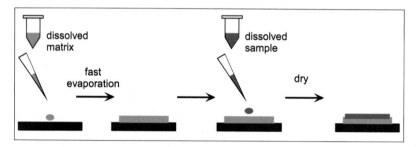

Fig. 4.8. The principle of the fast evaporation method.

A homogenous mixture is essential for obtaining good spectra. A droplet with a volume of 0.5 to 1 µL is placed onto the stainless steel target plate and dried by ambient pressure evaporation, heating with a stream of warm air or under vacuum until crystallisation occurs.

For the *fast evaporation method* (Fig. 4.8), a water-insoluble matrix is used. The matrix is dissolved in an organic solvent like acetone and a drop is applied to the target plate. The solvent evaporates within a few seconds leaving a dry thin film of the matrix on the target. A drop of analyte solution is then applied on top of the dried matrix. The analyte molecules are absorbed into the matrix crystal close to the matrix surface. It is possible to wash the crystal with water several times to remove impurities, especially alkaline metal ions from buffers. With the fast evaporation method, often spectra of high sensitivity and high resolution can be obtained.

4.2.6   *Applications of MALDI*

MALDI is mainly used for the analysis of proteins and peptides and their mixtures. It is possible to determine the molecular weights, to obtain structural information and to investigate post-translational processes. Molecular weights of proteins and peptides can be determined accurately with only a small amount of sample. The protein structure can often be determined by digestion with an enzyme and analysing the obtained characteristic peptide fragments. Changes in the protein structure and post-translational processes such as the formation of sulfide bonds or glycosylation can also be identified and localised with MALDI-MS techniques.

The strong points of MALDI include the very *low amount of sample* necessary for analysis, a few fmol are sufficient. Unlike other ionisation methods, MALDI *tolerates* moderate concentrations of *buffer and salts* in the analyte mixture. Sample *preparation is relatively easy* and the spectra obtained are simple, so that even *mixtures can be analysed* without the need to separate the components prior to MALDI analysis. However, in contrary to ESI, MALDI cannot be directly coupled to liquid chromatography (LC) or capillary electrophoresis (CE) as it is not a continuous but a batch ionisation method.

A typical MALDI spectrum is shown in Fig. 4.9. It is the spectrum of r-hirudin, a protein consisting of 64 amino acids. As little or no fragmentation occurs, only one *major peak*, $[M+H]^+$, and two minor peaks, $[M+2H]^{2+}$ and the agglomerate $[2M+H]^+$ are observed. This peak-pattern is very typical of MALDI-TOF spectra. Depending on the salt concentrations in the buffers used for sample preparation, peaks like $[M + Na]^+$ and $[M + K]^+$ are also observed.

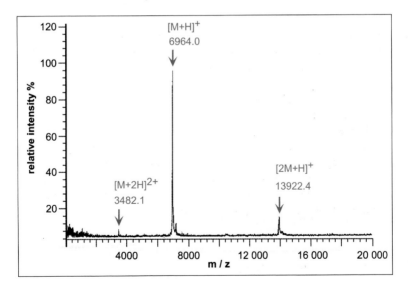

Fig. 4.9. Positive ion mass spectrum of the protein r-hirudin with MW 6963.5 Da. The matrix used was sinapinic acid (courtesy of Olaf Börnsen, Novartis).

The spectrum in Fig. 4.9 demonstrates that MALDI-TOF is a powerful method for accurate *molecular weight determination* of peptides and proteins. As there is almost no fragmentation, *mixtures* of peptides and proteins can be analysed without having to separate the compounds prior to analysis. In this respect, MALDI-TOF has to be regarded as a very fast separation method and is in many ways more powerful than chromatography or electrophoresis. In Fig. 4.10, a spectrum of low fat bovine milk is shown. The milk sample was added to the matrix without any pre-treatment and the different components present in the sample are resolved in the obtained mass spectrum.

Identification of proteins can also be achieved with MALDI-TOF by measuring a *"peptide fingerprint"* of the protein and comparing it to a database. The protein, usually taken from a 2D gel electrophoresis-plate (see section 3.2.5) is reacted with an enzyme, which cleaves the amino acid chain in specific places (Fig. 4.11). For example, trypsin, the most commonly used enzyme, cleaves the protein after Lysines and Arginines resulting in peptide fragments of several hundred to several thousand Da (see section 7.6.1). This fragment mixture, the peptide fingerprint, is very specific for a given protein (Fig. 4.12). Data obtained from the MALDI spectrum can be compared to a database containing theoretically calculated fin-gerprints for thousands of proteins. Often, the protein of interest can be identified unambiguously.

MALDI is more reliable for protein identification than other, commonly used methods, such as identification due to migration patterns in 2D gel electrophoresis

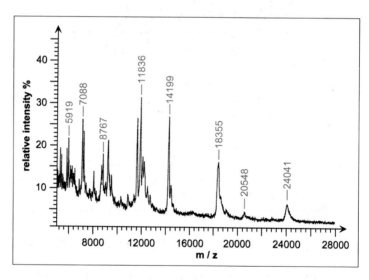

Fig. 4.10. MALDI spectrum of low fat bovine milk (courtesy of Olaf Börnsen, Novartis).

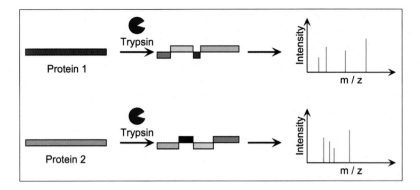

Fig. 4.11. Principle of protein digestion with trypsin to obtain a peptide fingerprint.

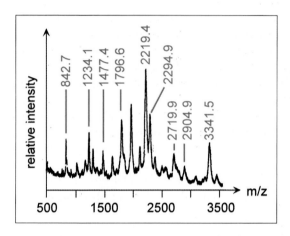

Fig. 4.12. Example of a peptide fingerprint: a tryptic digest of bovine serum albumin (BSA); (redrawn with permission, W.J. Henzel et al., PNAS 1993, 90: 5011–5015, copyright 1993, National Academy of Sciences, U.S.A.).

(section 3.2.5) and retention times in liquid chromatography (chapter 2). MALDI can, however, only be used, if the protein of interest is already known and kept in a database. If the protein is completely unknown, de novo sequencing of the amino acid chain becomes necessary (see chapter 6.3).

4.3 Electrospray Ionisation Mass Spectrometry (ESI-MS)

Electrosprays are generated by dispersing a liquid into small droplets via an electric field. This method has been known for a long time and is used for a variety of tasks ranging from metal spraypainting to ionisation of samples in mass spectrometry. First experiments on electrospray ionisation (ESI) of polymers were undertaken by Malcolm Dole in the late 1960s. ESI for mass spectrometry as used in modern instruments today was developed by John Fenn in the 1980s. In 2002, Fenn was awarded the Nobel Prize for his invention.

ESI enables the production of molecular ions directly from samples in solution. It can be used for small as well as large biopolymers up to about 200,000 Da including peptides, proteins, carbohydrates, DNA fragments and lipids. Unlike MALDI, ESI is a *continuous ionisation method* and suitable for coupling with liquid separation methods like HPLC (chapter 2) or CE (chapter 3.3).

4.3.1 *Ionisation Principle*

Electrospray ionisation is based on the dispersion of a liquid with the help of an electric field. The sample solution, containing analyte ions, is pumped into a heated chamber through a capillary or needle. A potential difference of several kilovolts is applied between the capillary and the opposing chamber wall (Fig. 4.13), creating an intense electric field at the capillary exit. If the capillary has a positive potential, negative ions are held back and positive ions are drawn away from the capillary towards the opposing chamber wall. This leads to the formation of a liquid cone at the end of the capillary. Droplets with positively charged analyte ions form at the tip of this cone. These are dragged through the chamber by the electric field whilst continuously loosing solvent due to evaporation. The droplets shrink which leads to an increase of charge density on the droplet surface. The repulsive forces on the droplet surface move eventually so close together, that the droplet bursts into

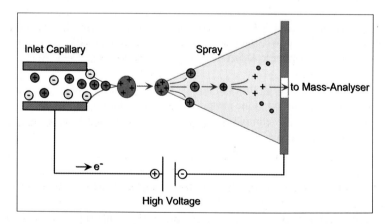

Fig. 4.13. The principle of electrospray ionisation.

a mist of finer droplets. This process of shrinking and bursting occurs repeatedly until, eventually, the analyte is completely desolvated and transferred into the mass analyser.

A typical feature of electrospray ionisation is the formation of multiply charged ions. For larger biomolecules a series of signals is obtained, consisting of $[M + H]^+$, $[M + 2H]^{2+}$, $[M + 3H]^{3+}$ to $[M + nH]^{n+}$ signals. As these highly charged ions appear at relatively low m/z values in the mass spectrum, ESI-MS allows observation of very high molecular weights, which are not accessible by other techniques. Samples suitable for ESI have to be soluble and stable in solution and need to be relatively clean. Ion formation in the spray is hindered by buffers, salts and detergents. These have to be kept to an absolute minimum.

The potential difference between the capillary and the channel wall can be applied in two ways, depending on whether cations or anions are to be analysed. (1) In the *positive ion mode*, the capillary has a positive potential. Negatively charged ions are held back by the capillary and cations are dragged through the chamber into the mass analyser and detected. Often low pH values of the sample solution are used to promote formation of cations. (2) In the *negative ion mode*, the potential difference is reversed – the capillary is negative. Cations are held back whereas anions are drawn towards the analyser. In this mode, high pH values, pH > pI, are employed.

4.3.2   *ESI-Source and Interface*

Electrospray ionisation is achieved at atmospheric pressure, the mass analyser, however, operates under high vacuum. A special *interface is* therefore necessary to transfer the ions from the ionisation chamber into the mass spectrometer. A schematic of such an interface is shown in Fig. 4.14. Usually a zone of intermediate pressure separates the ionisation chamber and the mass analyser. The liquid sample together with a *curtain* or *nebulising gas* is introduced into the heated ionisation chamber. An electrospray is generated by applying a potential difference between the needle and the opposite interface plate. A small proportion of the desolvated analyte ions exit the ionisation chamber through a submillimeter orifice and enter the zone of intermediate pressure. The analyte ions then pass via another small orifice into the mass analyser. This is usually a quadrupole which is operated under high vacuum.

A characteristic feature of ESI is that the sample can be pumped into the mass analyser *continuously*. MALDI, on the other hand, is a pulsed method which requires a dry sample. Thus, ESI-MS can be coupled directly to liquid separation methods such as RP-HPLC (section 2.3.1) and CE (section 3.3). As the sample emerges from the separation column it is directly pumped into the electrospray chamber. As outlined earlier, MALDI-TOF is capable of separating

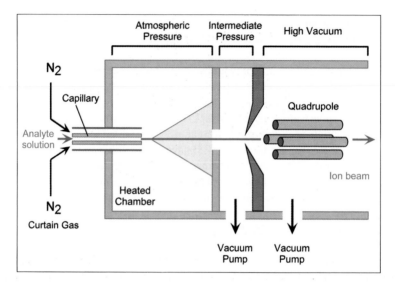

Fig. 4.14. ESI interface, connecting the ionisation chamber at atmospheric pressure to the mass analyser under high vacuum.

sample components directly from the sample mixture (Fig. 4.10); ESI-MS has to be coupled to LC or CE for separation of sample components.

Depending on the amount of sample available, different *flow rates* are used. A low flow rate allows for long measurement times to optimise instrument parameters. With the *pneumatically assisted electrosprays* as shown in Fig. 4.13, rather large capillaries of 50–100 μm and flow rates of 5–200 μL min^{-1} are used. In *micro-electrospray*, capillaries with 10–25 μm diameter and flow rates of 0.2–1 μL min^{-1} are employed. For bioanalysis often only a limited amount of sample is available, requiring very low flow rates in the nanolitre per minute range. *Nano-electrosprays* can be operated at 5–20 nL min^{-1} by using 3–5 μm diameter capillaries.

4.3.3   *Quadrupole Analyser*

The mass analyser most commonly used with ESI is the *quadrupole analyser*. The quadrupole is essentially a mass filter. At a given set of parameters only ions with a specific *m/z* value pass through the quadrupole and reach the detector. By scanning over an *m/z* range, whole spectra can be obtained.

The quadrupole analyser consists of four parallel rod-like metal electrodes. A direct current (DC) and an alternating current (AC) field are applied to these electrodes (Fig. 4.15). At a given field, ions of one defined *m/z* ratio can pass

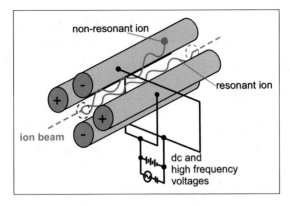

Fig. 4.15. Principle of the quadrupole analyser.

through the quadrupole following an oscillating pathway. These resonant ions reach the detector. All other ions are non-resonant and stopped by the quadrupole. To obtain a spectrum with all *m*/*z* ratios, the potential of the direct current and the amplitude of the alternate current are increased and all ions sequentially become resonant and reach the detector. This explains the low sensitivity of quadrupole analysers in comparison to time-of-flight analysers. In a quadrupole analyser, only a very small proportion (<1%) of the ions reaches the detector. It can be shown that the ion-mass is proportional to the potential of the dc and the amplitude of the ac. Thus, a mass spectrum can be obtained directly from the electrical field values. Quadrupole analysers are said to be easy to handle and robust. They can measure up to *m*/*z* values of 4,000, and resolutions between $R_S = 500$ and $R_S = 5,000$ can be achieved.

4.3.4 *Applications of ESI-MS*

ESI is suitable for almost all kinds of biomolecules, as long as they are polar and soluble in a solvent system that can be used for spraying. Peptides, proteins, carbohydrates, DNA fragments and lipids are all commonly analysed via ESI-MS. Molecular weight determination is one of the main applications. Furthermore, sequencing of peptides and DNA fragments (section 6.3) is possible with ESI connected to a *tandem mass spectrometer* (ESI-MS/MS).

ESI is a soft ionisation technique capable of ionising large biomolecules with little to no fragmentation; even non-covalent complexes remain intact and can be analysed. Fragmentation, if desired, can be controlled by changing the spray settings. As mentioned earlier, ESI can readily be coupled to liquid separation methods such as chromatography and capillary electrophoresis.

A problem with electrospray ionisation is its low tolerance for impurities or additives. Buffer and salt concentrations of more than 0.1 mM can prevent sufficient ion formation in the electrospray process, as can certain detergents at concentrations of more than 10 μM. Buffers commonly used in bioanalysis contain 100 mM phosphate and 150 mM NaCl and are thus unsuitable for ESI-MS.

Volatile organic solvents such as methanol, ethanol and acetonitrile are typically contained in the sample solution for electro spraying. Sometimes a volatile organic acid such as formic acid is added to promote cation formation. In almost every case it is necessary to clean the sample from salt contents and impurities prior to introduction into the electrospray chamber. Commonly used techniques for desalting include microdialysis and solid phase microextraction, which are quite labour intensive. Reversed phase liquid chromatography (RP-LC) (section 2.3.1) can be used for preconcentrating and isolating the sample compounds of interest. It can be coupled directly to ESI-MS as the organic solvents used in RP-LC are compatible with electrospray ionisation. At low flow rates, the sample can be injected directly from the column into the ionisation chamber; at higher flow rates the sample stream is split and only a fraction is directed into the mass spectrometer.

The ESI-MS spectrum of neurotensin, a peptide consisting of 13 amino acids with a molecular weight of 1,672 Da is shown in Fig. 4.16. Due to the soft ionisation, no fragments are observed. As mentioned earlier, ESI promotes the formation of multiply charged ions. Peptides and proteins, thus, give a series of signals with $[M + H]^+$, $[M + 2H]^{2+}$, $[M + 3H]^{3+}$ to $[M + nH]^{n+}$.

The number of peaks depends on the size of the molecule as well as the number of acidic and basic groups. Larger proteins can have a signal series with ions of up to $[M + 100H]^{100+}$. Isotopes are detected in addition to these peaks, leading to an overall rather complex spectrum with a large number of signals. How is it possible to determine which peak refers to the $[M + H]^+$ ion and, thus, the molecular

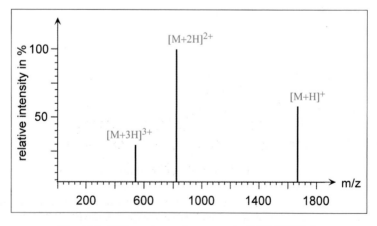

Fig. 4.16. ESI-spectrum of neurotensin (MW 1672 Da).

weight of the analyte molecule? If the resolution is good enough to see different isotopes, these can be used. In case of singly charged ions, the difference between the isotope peaks is exactly 1 Da, whereas for doubly charged ions, the difference between isotope peaks is only 0.5 Da. Often, the isotopes cannot be resolved, as in the spectrum shown in Fig. 4.17. The molecular weight is then calculated by software algorithms included within the instrument software. These algorithms produce a so-called *deconvoluted spectrum* (Fig. 4.18), that give the molecular weight in the form of a peak.

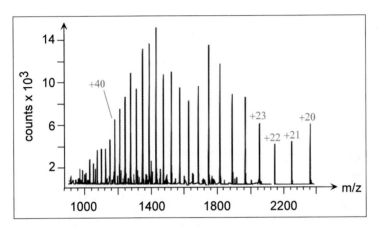

Fig. 4.17. ESI-MS spectrum of lactose permease showing differently ions carrying 20 to more than 50 positive charges (redrawn with permission from J.P. Whitelegge et al. PNAS, 1999, 96: 10695–10698, copyright 1999, National Academy of Sciences, U.S.A.).

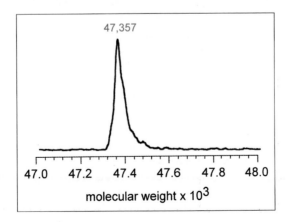

Fig. 4.18. The deconvoluted spectrum of lactose permease showing a molecular weight of 47,357 Da (redrawn with permission from J.P. Whitelegge et al. PNAS, 1999, 96: 10695–10698, copyright 1999, National Academy of Sciences, U.S.A.).

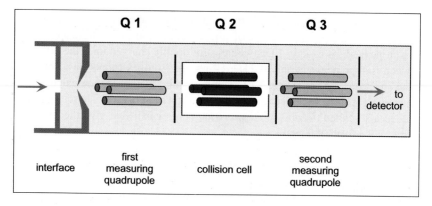

Fig. 4.19. Schematic of a tandem mass spectrometer.

Structural analysis of peptides (chapter 7), nucleic acids (chapter 6.3) and oligosaccharides can be performed with a *tandem mass spectrometer*, for example an ESI-MS/MS. In such a tandem mass spectrometer, three quadrupoles are arranged in series (Fig. 4.19). The first quadrupole ($Q1$) is a measuring quadrupole for determining the m/z of the introduced sample. The second quadrupole ($Q2$) acts as a reaction zone. It is a cell filled with an inert gas such as nitrogen, helium or argon. The analyte ions collide with the gas molecules and become fragmented, a process called *collision induced dissociation (CID)*. These fragments are then introduced into the third quadrupole ($Q3$) for mass analysis.

The measuring quadrupoles can be run in a *static* or *scanning mode*. In the *static mode*, the electric fields are kept constant and only ions with one defined m/z-value can pass. In the *scanning mode*, the quadruple sequentially allows ions within a defined m/z range to pass through. With these two options and two measuring quadrupoles, several modes of operation are possible (Table 4.2). The *daughter ion analysis* method is the most commonly used approach. For this method, the first quadrupole is set in a static mode and only ions with one specific m/z-value can pass. They are then reacted and fragmented in the collision cell. Ions resulting from this fragmentation, the so-called daughter ions are then analysed in the third quadrupole.

Tandem mass spectrometry can be applied for analysis of peptide mixtures. The first quadrupole only passes one specific peptide ion, which is then fragmented in the collision chamber, i.e. amino acids are cleaved from the peptide chain. In the third quadrupole, the difference between mass peaks gives information about the amino acid sequence in the peptide. An example of peptide sequencing with ESI-MS/MS is shown in Fig. 4.20. Oligonucleotides and oligosaccharides can be analysed in a similar fashion.

Table 4.2. Modes of operation for tandem mass spectrometry.

Experiment	Mode of $Q1$	Mode of $Q3$
daughter ion analysis	static (parent mass selection)	scanning
parent ion analysis	scanning	static (daughter mass selection)
multiple reaction monitoring	static (parent mass selection)	static (daughter mass selection)
constant neutral loss spectrum	scanning (synchronised with $Q3$)	scanning (synchronised with $Q1$)

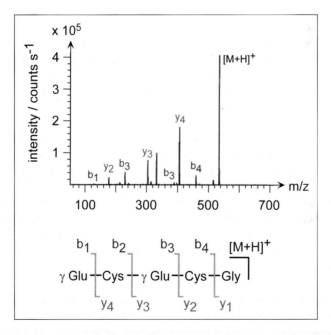

Fig. 4.20. Peptide sequencing of phytochelatin with ESI-MS/MS (V. Vacchina et al., *Analyst*, 1999, 124: 1425–1430; reproduced with permission of the Royal Society of Chemistry).

Summary

MALDI and ESI mass spectrometry are both powerful techniques for the analysis of high molecular weight biomolecules (Table 4.3) with applications including high accuracy determination of molecular weights, fingerprinting of peptides and structural analysis of peptides, oligonucleotides and oligosaccharides.

Table 4.3. Comparison of MALDI and ESI.

	MALDI	ESI
analysis of mixtures	possible	pure compound required
limit of detection	very low	higher due to losses in spray and analyser
coupling	not possible	possible to LC and CE
fragmentation	no fragments	some fragments, depending on applied voltages

For MALDI, samples are co-crystallised with a matrix and desorbed by laser pulses. The desorption process is very mild and spectra contain hardly any fragments. The method allows the analysis of relatively crude samples with very low limits of detection. Coupling to liquid separation methods is not possible; however, the time of flight analyser separates ions according to their *m/z* ratio in microseconds and thus allows analysis of mixtures without any sample pretreatment.

In ESI, the sample is dissolved in a volatile solvent. Sample pretreatment is more labour-intensive as impurities and salt concentrations have to be kept to a minimum. Detection limits are not as low as with MALDI due to the loss of sample during the electrospraying process as well as in the quadrupole analyser. Multiply charged molecular ions allow molecular weight determination of very large biomolecules. Fragments can be observed and enable sequencing and identification. Because ESI is a continuous ionisation method, direct coupling to chromatography and electrophoresis is possible.

References

1. G. Siuzdak, *Mass spectrometry for Biotechnology*, Academic Press, 1996.
2. R. A. W. Johnstone, Malcolm E. Rose, *Mass Spectrometry for Chemists and Biochemists*, 2nd edition, Cambridge University Press, 1996.
3. J. R. Chapman, *Mass Spectrometry of Proteins and Peptides*, Humana Press, 2000.
4. E. De Hoffmann and V. Stroobant, *Mass Spectrometry: Principle and Applications*, 2nd edition, John Wiley and Sons, 2001.

Chapter 5

MOLECULAR RECOGNITION
BIOASSAYS, BIOSENSORS, DNA-ARRAYS AND PYROSEQUENCING

In this chapter, you will learn about...

♦ ...bioassays which make use of the specific molecular recognition between antibodies and antigens.

♦ ...bioassays that are employed in clinical diagnostics for pregnancy tests and testing for HIV antibodies.

♦ ...the concept of biosensors with focus on the glucose biosensor.

♦ ...molecular recognition between single strands of DNA and how it is used in genomics and proteomics for DNA arrays.

♦ ...determination of a nucleotide sequence with pyrosequencing and how single nucleotide polymorphisms (SNPs) can be detected.

The underlying principle of the analytical methods described in this chapter is that of *biomolecular recognition*; the ability of a biomolecule to interact with one other particular type of biomolecule, like a key fitting a lock. With this key-lock principle, it is possible to specifically detect the target molecule, which could be an antigen, an antibody, a hormone, a DNA fragment or a sugar, even in very complex sample mixtures like urine or untreated blood.

In bioassays, the molecular recognition between antibody and antigen as well as the signal generation for detection usually occur in solution or on an inert solid phase, whereas in biosensors, they are closely integrated on the surface of an active electronic device. DNA-arrays make use of the specific recognition of single strands of DNA, which only combine to form a double strand where there is a perfect match. Single nucleotide polymorphism (SNP) detection with pyrosequencing also relies on the specific base pairing of A-T and G-C.

The general principles of these techniques will be shown in the following sections and manifested by examples.

5.1   Bioassays

Bioassays, most importantly *immunoassays*, rely on the highly specific reaction between antibody and antigen – the fundamental molecules of the immune system. Detection limits as low as in the order of fmol can be achieved for some assays. Bioassays are one of the most common methods in bioanalytical chemistry, especially for diagnosis and management of diseases.

Antibody (Ab) and *antigen* (Ag) both feature recognition sites, called *paratope* and *epitope*, respectively. If the paratope of the antibody matches the epitope of the antigen, an *Ab–Ag complex* is formed (Fig. 5.1). An antibody only reacts with a matching antigen and with this specific antigen only, other antigens in the reagent mixture remain unbound. In many cases, this high *specificity* enables the direct analysis of complex sample mixtures such as untreated blood or urine. The *affinity* of antibody to antigen is very high and binding occurs even at very low concentrations. This explains the high *sensitivity* and low limits of detection obtained with bioassays.

To detect the assay product, it is usually necessary to use a *label*, which is attached either to the antibody or the antigen. This label can be fluorescent, luminescent, radioactive, an enzyme or an electrochemically active group. Immunoassay reactions can be performed in a large variety of formats, in solution or on a solid support, with limited reagent or an excess of reagent. These formats are discussed in more detail in the following sections, after a description of antibody and antigen structure and immunocomplex formation.

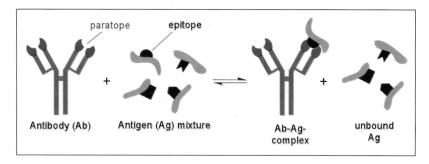

Fig. 5.1.   Molecular recognition of antibody (Ab) and antigen (Ag) resulting in an Ab–Ag-complex and unbound Ag.

5.1.1 *Antibodies*

Antibodies are a major class of soluble proteins, they constitute about 20 % of the total plasma protein. When a foreign substance, a so-called antigen, enters the body of a human or an animal, the immune system responds by producing antibodies in large quantities. These antibodies bind specifically to the antigen to form an immunochemical complex and, thus, help to eliminate the foreign substance from the body. Unlike enzymes, antibodies do not act as catalysts for a reaction; they solely bind to the antigen with very high affinity. To produce antibodies for laboratory use, hosts are injected with the appropriate antigen. Mice, goats, chickens, rats, rabbits, horses, donkeys, hamsters and humans are common hosts for antibody-production. Alternatively, antibodies can be produced *in vitro* with specialised cell cultures.

Antibodies, also referred to as *immunoglobulins* (Ig), consist of four subunits: two identical *light chains* (*L*), with a molecular weight of about 25 kDa and two identical *heavy chains (H)* with a molecular weight of about 50 kDa. These subunits are associated via disulfide bonds and non-covalent interactions to form a *Y*-shaped symmetric dimer $(L-H)_2$ (Fig. 5.2).

There are five classes of immunoglobulins: IgA, IgD, IgE, IgG and IgM. These classes are determined by the five different types of heavy chains. There are also two types of light chain and these can appear in any of the five Ig classes. Depending on their class, immunoglobulins fulfil different physiological functions. Immunoglobulin G (IgG) is the most common antibody in the body with an abundance of about 70 %.

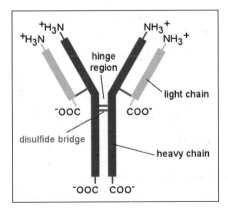

Fig. 5.2. A Y-shaped immunoglobulin consisting of two identical heavy chains (*H*) and two identical light chains (*L*) connected via disulfide bridges.

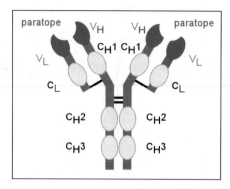

Fig. 5.3. Constant (C) and variable (V) regions within the immunoglobulin molecule. The variable regions at the N-terminal of the light and heavy chains make up the paratopes of the antibody.

Within the immunoglobulin molecule, there are *constant* (C) and *variable* (V) *regions* consisting of a specific folding pattern of the amino acid sequence (Fig. 5.3). The constant regions are the same for every antibody of that class, for example all IgG molecules have the same constant regions. One of these constant regions is located at the C-terminal of the light chain (C_L) and three further constant regions are located at the C-terminal of the heavy chain (C_H1, C_H2, C_H3). The variable regions make up the paratopes of the antibody. These regions target the antigen of interest. The paratopes are located at the tips of the Y-shape at the N-terminal of the light chains and heavy chains and are called V_L and V_H, respectively.

As can be seen from Fig. 5.3, each antibody molecule has two identical binding sites (paratopes) for the antigen. Thus, the antibody is *bivalent*. The size and shape of the crevice depends on the amino acid sequence within the V_L and V_H region.

Two types of antibodies can be distinguished: *monoclonal* and *polyclonal* antibodies. Polyclonal antibodies are essentially a mixture of antibodies as produced by a host upon injection of an antigen. They can bind to several epitopes on the antigen. Monoclonal antibodies on the other hand bind to only one particular epitope on the antigen. They are more specific and reproducible and thus usually preferred for analytical assays. Such monoclonal antibodies have to be produced from cell cultures.

When treated with enzymes, immunoglobulins can be cleaved into fragments. The enzyme *papain* cleaves IgG into three fragments of about 50 kDa (Fig. 5.4). The two identical *Fab*-fragments originate from the arms of the Y-shape, the N-termini of the heavy and light chains. As the Fab-fragments contain the paratopes, they retain the antigen binding ability (ab = antigen binding). The *Fc*-fragment originates from the stem of the Y-shape and contains the C-termini

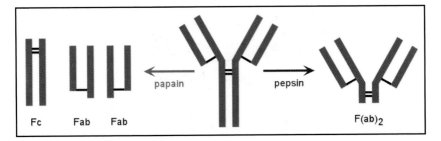

Fig. 5.4. Cleaving of immunoglobulin by papain, resulting in two identical Fab-fragments and one *Fc*-fragments and by pepsin, resulting in an $F(ab)_2$-fragment.

of the heavy chains linked via disulfide bridges. The *Fc*-fragment can be easily crystallised (c = crystallisable). It does not have any antigen recognition sites but retains some other antibody functionalities. The enzyme *pepsin* cleaves an immunoglobulin at the stem below the hinge-region, resulting in an $F(ab)_2$ fragment with the arms of the Y still being joined. This $F(ab)_2$ fragment contains both paratopes. Occasionally, the *Fab* or $F(ab)_2$ fragments are used in immunoassays instead of the whole immunoglobulin molecule.

5.1.2 *Antigens*

An antigen is a molecule capable of inducing an immune response when entering the body. Two classes can be distinguished: *complete* and *incomplete antigens*. Complete or full antigens induce an immune response by themselves. They are usually large proteins like albumin (66 kDa) or ferritin (580 kDa). These full antigens can have several copies of the same epitope or they can be *multi-determinant*, i.e. they contain several different epitopes that bind to different antibodies. For example, an antigen with three different epitopes can stimulate the production of three different antibody molecules with different paratopes. Incomplete antigens, also called *haptens*, are lower molecular weight molecules like the drug theophylline (180 Da) or the steroid hormone *cortisone* (362 Da). They cannot induce an immune response by themselves, i.e. they are not *antigenic* by themselves. However, if attached to a protein carrier, the production of specific antibodies against these haptens can be triggered. Once produced, these anti-hapten antibodies will recognise the hapten even without the protein carrier. Haptens usually only feature a single epitope. If conjugated to a suitable carrier, virtually any chemical substance can serve as an incomplete antigen.

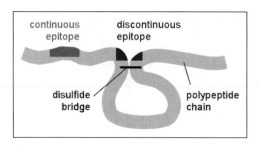

Fig. 5.5. Continuous and discontinuous epitope in the polypeptide chain of an antigen molecule.

The binding site of the antigen, the epitope, makes up only a small area of the total antigen structure; usually not more than 18 amino acids of the antigen interact with the paratope of the antibody. Epitopes can be *continuous* or *discontinuous* (Fig. 5.5). Continuous or linear epitopes are common in fibrous proteins. Conformational or discontinuous epitopes on the other hand are generated through folding. Such epitopes are associated with globular proteins and helical structures and can potentially be destroyed upon denaturation, for example, if disulfide bonds are split. Antibodies are capable of distinguishing between antigens even if they are chemically very similar. It is the overall three-dimensional structure of the antigen rather than the specific chemical property that defines its affinity and interaction with an antibody.

5.1.3 *Antibody-Antigen Complex Formation*

In the case of matching paratope and epitope, antibody and antigen form a complex, also called *immuno-complex* (see Fig. 5.1). This complex formation is reversible and depends on the interplay of several forces. When antibody and antigen approach each other in the blood stream or in a reaction vessel, the primary binding force acting over a relatively long range of about 10 nm is the *electrostatic interaction* between the positively charged amino groups and the negatively charged carboxyl groups of the polypeptide molecules. At closer range, exclusion of hydration water molecules can occur and this leads to the formation of *hydrogen bridges* between hydroxyl, amino and carboxyl groups. Finally, at very close range, *van der Waals-forces* come into play. These involve the interaction between external electronic clouds and induced dipole moments. Additionally, non-polar groups can associate with each other in an aqueous environment leading to hydrophobic interactions.

The binding strength between a single epitope and paratope is referred to as their *affinity*. It is dependent on the number and strength of the bonds formed between the epitope and paratope. Affinity can be quantified by the equilibrium constant (K_{eq}) for the complex formation:

$$K_{eq} = \frac{[Ab - Ag]}{[Ab][Ag]} \qquad \text{(equation 5.1)}$$

with $[Ab]$, $[Ag]$ and $[Ab - Ag]$ being the concentration of antibody, antigen and immuno-complex, respectively. K_{eq} is usually in order of 10^6 to 10^{12} L mol^{-1}. However, for bioassays only antibody–antigen pairs with equilibrium constants of larger than 10^8 L mol^{-1} are used. Affinity must be differentiated from *avidity*, which describes the interaction of antigens with multiple epitopes with antibodies with more than one paratope.

Although binding between antibody and antigen is highly specific, in practice there are often problems due to *non-specific bonding* of proteins to the walls of the reaction vessel or to each other, usually caused by association of opposite charges. This non-specific bonding can be minimised by adjusting the pH of the buffer, addition of ions in the form of salts or the addition of surfactants such as sodium dodecyl disulfate (SDS) or polymers such as polyethylene glycol (PEG).

5.1.4 *Assay Formats*

Immunoassay reactions can be performed in a large variety of formats. They can be classified according to three main criteria: (1) limited or excess reagent, (2) homogeneous or heterogeneous and (3) labelled or unlabelled.

In *limited reagent* assays or *competitive* assays, a limited amount of antibody is used, which is insufficient to bind with all the antigen molecules of the assay. The sample containing an unknown amount of antigen is mixed with a fixed and known amount of labelled antigen (Fig. 5.6). Unlabelled sample antigen and labelled antigen compete for the limited number of antibody binding sites. The concentration of unlabelled antigen can be determined from the proportion of labelled antigen that is bound to the antibody or from that which remains free. Usually, it is necessary to separate the bound and free fractions. Immobilising the limited reagent onto an inert solid surface facilitates this separation step. To obtain quantitative results, it is essential to keep the ratio of limited antibody reagent to added labelled antigen constant.

The assay design must be altered according to whether the sample is an antigen or antibody. For antigen analysis, as described above (Fig. 5.6), the antibody is the limited reagent. It is incubated with a mixture of sample antigen and a known amount of labelled antigen. If, on the other hand, the sample is an antibody, the format is reversed. The antigen is then the limited reagent, and this is incubated with a mixture of sample antibody and labelled antibody (Fig. 5.7).

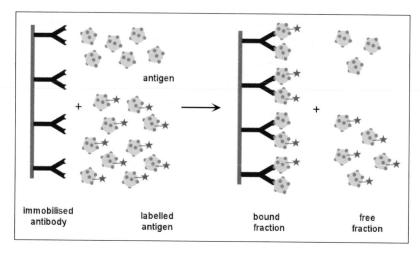

Fig. 5.6. Principle of competitive immunoassays: the antigen molecules of the sample compete with a fixed amount of labelled antigen for the limited amount of antibody binding sites.

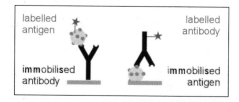

Fig. 5.7. Assay design for antigen sample (left) and antibody sample (right).

The competitive assay format is very sensitive when small amounts of sample molecules are used. For example, in a format with antibody as the limited reagent, the smaller the concentration, c, of antigen molecules in the sample, the more labelled antigen molecules will bind to the limited amount of antibody, resulting in a large signal. A *dose-response curve* for a competitive assay is shown in Fig. 5.8. Within a certain range of c, the signal intensity, s, is inversely proportional to the analyte concentration: $\Delta c \propto -\Delta s$. For very high and very low analyte concentrations the curve flattens out and quantitative analysis is not possible anymore. In assay formats other than the limited reagent format, the signal becomes smaller as the number of sample molecules decreases. Competitive immunoassays can be used for both large analytes with several binding sites and small analytes with potentially only one binding site.

In the *excess reagent* or *non-competitive* format, the antigen sample is incubated with an excess of antibody reagent (Fig. 5.9). All the antigen molecules bind, but

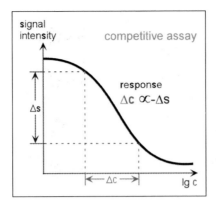

Fig. 5.8. Response curve for a competitive assay format. The smaller the sample concentration, the larger the signal generated.

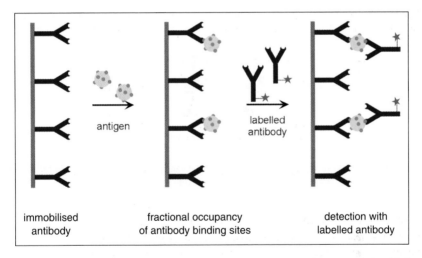

Fig. 5.9. The principle of a non-competitive immunoassay: the antigen sample is added to an excess of antibody reagent leading to fractional occupancy of antibody binding sites. A secondary antibody with a label is then added and a sandwich complex is formed allowing detection.

not all the antibody binding sites are occupied. To detect the amount of antigen attached to an antibody, a second, labelled antibody is added which binds to another epitope of the antigen. This leads to the formation of a *sandwich-complex* between the *primary antibody*, the sample antigen and the labelled *secondary antibody*. After washing off any excess reagent, the sandwich complexes containing the label can be detected and the signal generated is directly related to the amount of antigen in the sample. A sandwich immunoassay is better suited for large analyte molecules which are likely to have several epitopes. For sandwich-complex formation at least two binding sites are required on the analyte molecule.

This method can be modified for the detection of antibodies by immobilising an excess of suitable antigens onto a substrate. These antigens are incubated with the primary antibody. Then a secondary, labelled antibody is added which binds to the primary sample antibody, usually to the *Fc* region of the primary antibody (Fig. 5.10). After washing away the unbound secondary antibody, the signal from the label can be detected and directly correlated to the amount of primary antibody. This method is used for HIV antibody detection and is discussed in more detail later in this chapter.

The response curve in case of a non-competitive assay is shown in Fig. 5.11. The labelled secondary antibody binds directly to the sample molecules. So the larger the number of sample molecules, the larger the signal: $\Delta c \propto \Delta s$. This is only true within a certain range of sample concentrations – for very high and very low

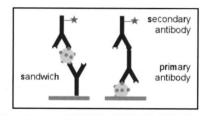

Fig. 5.10. Use of a secondary antibody as part of a sandwich with the antigen in the middle (left) and directed against the primary antibody (right).

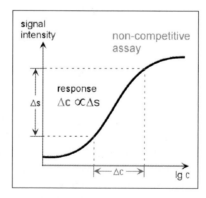

Fig. 5.11. Response curve for a non-competitive assay: the larger the analyte concentration, the larger the signal generated.

concentrations the response curve flattens out, similar to the curve for competitive assays (Fig. 5.8). In comparison to competitive assays, exact metering of reagents is less critical.

Heterogeneous assays require the separation of the antibody–antigen complex from the free unbound fraction before detection. Separation can for example be achieved by precipitation, coupling of antibody or antigen to a solid phase or by chromatographic techniques. However, this introduces a labour intensive step into the assay protocol and care has to be taken not to influence the antibody–antigen equilibrium. On the other hand, the separation and necessary washing steps also remove unreacted material for example from serum or urine samples. This purification can lead to an improvement in sensitivity.

Homogeneous assays can be performed without the need for separation. This can be advantageous with respect to convenience, time and cost of the assay and also facilitates automation. The detection method must be able to differentiate between bound and free antibodies without physically separating them from each other. An example of such a detection method is the turbidity measurement of the reaction mixture by laser light scattering or absorption. The more immuno-complexes are present in the reaction mixture, the higher its turbidity. Often, latex or gold particles are attached to the antibodies to enhance this turbidity effect.

Signal generation in almost all immunoassays requires attachment of a *label* to one of the reagents. This label can be (a) a radioisotope such as ^{125}I or ^{14}C, (b) an enzyme, for example, horseradish peroxidase or alkaline phosphatase, (c) a fluorophore such as fluorescein, (d) a luminescent species such as luminol or (e) micro- or nanoparticles made of latex or gold.

The method used for *detection* depends on the type of label used. Isotopic counting is employed for radioisotopes, colorimetry for enzyme assays, luminescence and fluorescence measurements can be achieved by means of photomultiplier tubes, while turbidimetry or nephelometry is used for particle enhanced assays.

As can be seen from the previous paragraphs, there is a wide variety of possible assay formats and labels, each with their own specific advantages and disadvantages. When designing a new immunoassay, it is essential to think about the requirements for the specific analyte. What limit of detection or sensitivity is required? What range of concentrations occurs in the samples and, hence, what is the required linear range of signal response? What is the nature of the sample, are washing steps essential? How many samples are to be measured and is automation desirable? Immunoassays for home-testing such as urine based pregnancy tests must be simple and robust enough to be used by a layperson without the need of specialised equipment. Biomedical laboratories on the other hand have trained personnel and dedicated equipment available, but require fast and auto-mated protocols. In the following sections, two examples of immunoassays with different formats are described in more detail – the home-pregnancy test and the HIV-antibody test.

5.1.5   *Home Pregnancy Test*

A few days after conception, the glycoprotein hormone *human chorionic gonadotropin (hCG)* appears in the urine and its concentration increases rapidly during the first weeks of pregnancy. Thus, hCG is an excellent marker for pregnancy. A number of immunoassays for home-testing of hCG in urine have been developed and form a significant part of the home-diagnostics market.

The assay is carried out on a test strip and based on a sandwich format with two antibodies. One antibody, the *capture antibody*, is immobilised, i.e. covalently attached to the device surface. The second antibody, the *tracer antibody*, is labelled, usually with a dye. This tracer antibody is impregnated onto the surface of the device, but is not permanently attached. The strip component is composed of an adsorbent material. Once the urine sample is applied, the liquid moves along the strip by capillary action and the assay-reactions are carried out in flow.

A schematic of a test strip is shown in Fig. 5.12. The adsorbent material is usually enclosed within a plastic casing featuring a sample input window, a test result window and a control window. A drop of urine is applied at the sample input and the liquid first moves over the zone, which contains the labelled tracer antibody (Fig. 5.13). If hCG is present in the sample, it forms a complex with the tracer antibody. This complex continues to move along the adsorbent material and passes over the area with the immobilised capture antibody. A sandwich is formed between the immobilised capture antibody and the tracer antibody with the hCG in between. Thus, the initially mobile antibody with the label becomes immobilised. The amount of sandwich complexes formed is directly proportional to the amount of hCG present in the sample. If the hCG concentration exceeds a minimum concentration, the dye colour becomes visible to the eye.

In Fig. 5.14, positive and negative test results are depicted for an assay using a red dye as a label. The control window always shows the colour of the label if the

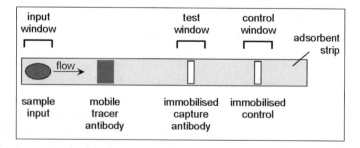

Fig. 5.12.   Schematic of a test strip device for an hCG immunoassay.

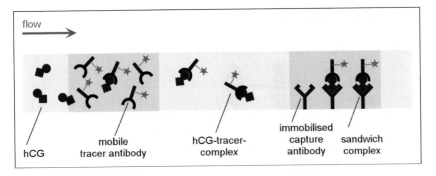

Fig. 5.13. Assay reaction in flow immunoassay.

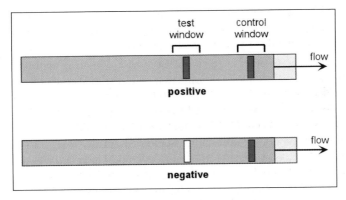

Fig. 5.14. Positive and negative results of a pregnancy test. The control window always shows the colour of the label, in this case a red dye. The test window only shows the colour of the label if the pregnancy marker hCG is present in the sample.

test has been carried out correctly. The test window only shows the dye colour if hCG is present in the sample.

Other assays commonly used for home-testing include cholesterol, blood-glucose, urine-glucose, blood type and alcohol-tests.

5.1.6 *Enzyme Immunoassays (EI and ELISA)*

Enzymes are one of the most commonly used labels in immunoassay. They enable measurement with a sensitivity close to that reached by radio-immunoassays but without the health hazards associated with radioactive substances. *Enzyme immunoassays (EI)* and most importantly *enzyme-linked immunosorbent*

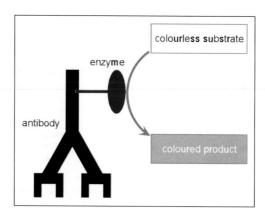

Fig. 5.15. Operating principle of an enzyme label: the enzyme acts as a catalyst for the conversion of a colourless substrate to a coloured product.

assays (ELISA) can be automated and have found use in many applications from monitoring drug levels in blood or urine to monitoring industrial processes.

In labelled fluorescence or isotopic immunoassays, the amount of bound analyte is directly related to the quantity of label that is present. This is not the case with enzyme immunoassays. Instead, the enzyme label acts as a catalyst for the conversion of a colourless substrate to a coloured product (Fig. 5.15). One single enzyme molecule can catalyse the conversion of a large number of substrate molecules and, thus, generate a large signal. This signal amplification is one of the main advantages of enzyme labels, making the quantitative analysis of low sample concentrations possible. For example, hormones in blood are often analysed with enzyme immunoassays. A disadvantage is the additional complexity since it is necessary to add substrate reagents and have a further reaction step.

5.1.6.1 *Enzymes*

Enzymes are high molecular mass proteins or glycoproteins that catalyse biochemical reactions. Essentially, they are bio-catalysts. An enzyme is highly specialised. It will only catalyse one particular reaction. The enzyme urease, for example, catalyses the hydrolysis of urea, whereas the enzyme DNA polymerase catalyses the synthesis of DNA. According to their activities, enzymes are classified into six groups (Table 5.1). Enzymes used in bio-analysis are often from group 1 (oxido-reductase) and group 3 (hydrogenase). For immunoassays, the most popular enzyme is horseradish peroxidase (HRP); although alkaline phosphatase (AP) and acetylcholine esterase (AChE) are also commonly used.

The enzyme-catalysed reaction from substrate to product usually proceeds in two steps (equation 5.2). First, the substrate (S) reversibly binds to a specific site on

Table 5.1. Classification of enzymes.

No.	Class	Catalysed reaction
1	oxido-reductase (dehydrogenase, oxidase, peroxidase, oxygenase)	oxidation-reduction
2	transferase	group transfer
3	hydrogenase	hydrolysis
4	lyase	bond cleavage
5	isomerase	isomerisation
6	ligase	bond formation

the enzyme (E) to form a complex (ES). In the second step the enzyme catalyses the conversion of the substrate to a product (P) (equation 5.2).

$$E + S \underset{k_2}{\overset{k_1}{\rightleftharpoons}} ES \overset{k_3}{\longrightarrow} E + P \qquad \text{(equation 5.2)}$$

The *turnover* of the enzyme, k_3, is a measure of how many molecules of substrate can be converted to the product within a period of time. The rate of the reaction, v, can be described by the *Michaelis-Menten equation*:

$$v = \frac{k_3[E][S]}{K_m + [S]} \qquad \text{(equation 5.3)}$$

The reaction rate depends on the concentrations of the enzyme $[E]$ and the substrate $[S]$, as well as on the turnover number, k_3, and the Michaelis-Menten-constant (K_m), which, broadly speaking, quantifies the enzyme's affinity for its substrate. The catalytic activity of an enzyme also depends on temperature, pH, ionic strength and the presence of inhibitors or activators. For a more detailed discussion of enzyme kinetics, refer to one of the biochemistry textbooks listed at the end of this chapter. A suitable enzyme for labelling must be stable under the necessary reaction conditions and it must have a high turnover rate.

The most popular enzyme label, horseradish peroxidase (HRP), catalyses the oxidation of hydrogen peroxide to water in the presence of a hydrogen donor (DH):

$$H_2O_2 + 2\,DH \overset{HRP}{\longrightarrow} 2\,H_2O + 2\,D \qquad \text{(equation 5.4)}$$

D and DH are the oxidised and reduced form of a hydrogen donor, respectively. This is usually a dye that changes colour when reduced, for example ABTS, 2,2'-azino-bis (ethyl-benzothiazoline-6-sulfonate), with an absorption maximum of $\lambda_{max} = 415\,nm$ for the reaction product.

Enzyme immunoassays are generally performed using a heterogeneous assay format on a microtitre plate containing typically 96 wells. These heterogeneous assays are referred to as enyzme-linked immunosorbent assays (ELISA). ELISAs are widely used in clinical testing. For example, one type of ELISA is currently used for the clinical screening of blood supplies for HIV, the Human Immunodeficiency Virus that causes acquired immunodeficiency syndrome (AIDS). This assay is discussed in more detail below as a typical example of an ELISA.

5.1.6.2   ELISA for HIV-Antibodies

The human body does not naturally produce HIV antibodies. They are only produced when an infection with the virus occurs. Hence, if antibodies are detected, this is a clear indication that HIV has entered the body.

The ELISA for HIV detection relies on the fact that HIV antibodies bind specifically to the antigenic virus (Fig. 5.16). First, HIV antigens are immobilised on the microtitre well surface. After a washing step, the sample, possibly containing HIV antibodies, is added and then left to incubate. After another washing step, a second antibody is added. This is targeted towards the HIV antibodies and contains an enzymatic label. This secondary antibody binds to the HIV antibodies, if they are present in the sample. Any unbound secondary antibody is then washed away. Finally, the appropriate enzyme substrates are added. If any secondary antibodies were bound then a colour reaction occurs between the substrates and the enzymatic labels of the secondary antibodies. The colour intensity can be measured with a spectrophotometer. It takes about one hour to perform this type of ELISA test.

Many different enzyme reactions can be used for immunoassays and the reaction product does not necessarily have to be a coloured dye. Reactions resulting in fluorescent, chemiluminescent or electrochemically active products can also be used.

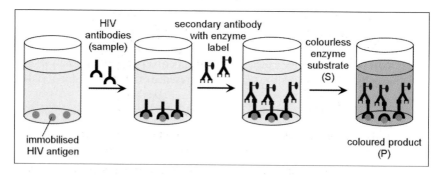

Fig. 5.16. Principle of an ELISA for HIV antibodies.

5.2 Biosensors

Although immunoassays can be used for the quantitative detection of extremely small amounts of analytes in complex samples mixture, they are often quite time consuming to perform. This is because several reaction and washing steps must be carried out before detection is possible. The biochemical reaction of the assay is generally separated from the measurement system. In *biosensors*, on the other hand, the two are intimately combined onto a single device without the need for additional reagents and washing steps:

$$biosensor = molecular\ recognition + signal\ transduction$$

The *bioreceptor* specifically recognises the target analyte (Fig. 5.17). This receptor can be almost any biological system that exhibits molecular recognition, for example, an antigen or antibody, an enzyme or even a whole cell. The bioreceptor is located in close proximity to the *transducer*. When a target molecule is present and a recognition event occurs, then the immediate environment of the bioreceptor changes. This change is converted into a measurable signal by the transducer.

These two components, the bioreceptor and transducer are integrated into one single sensor, and it is possible to measure the target analyte directly without using any additional reagents. For example, the glucose concentration in blood can be measured by just dipping the biosensor into the blood sample. Simplicity and speed of measurement are the main advantages of biosensors. Devices can be used by non-specialist operators at the point of care and this allows for immediate action to be taken. The development of a biosensor is, nonetheless, a challenge and so far only a few biosensors have become commercial products.

The biosensor-concept was first described by L.C. Clark and C.L. Lyons in the early 1960s. Since then, researchers have pin-pointed a wide variety of applications for qualitative and quantitative analysis (Table 5.2). Only a few biosensors have

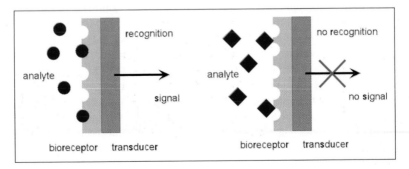

Fig. 5.17. A biosensor is comprised of a bioreceptor with a specific surface and a transducer to pass on a signal when recognition occurs.

Table 5.2. Applications for biosensors.

Field	Applications
health care	markers of diseases like myocardial infarction or cancer monitoring of administered drugs diagnosis of infectious diseases analysis of glucose and hormone levels
environmental	water and soil analysis pesticides and other toxic substances industrial effluent control
agriculture	pesticides, crop diseases
food control	food freshness determination of fruit ripeness by glucose content quantification of cholesterol in butter pathogenic organisms like *E. coli*
process control	fermentation monitoring
microbiology	bacterial and viral analysis

been made commercially available. An example of a widely used biosensor is the blood glucose sensor for home monitoring by diabetics.

5.2.1 *Bioreceptors*

A bioreceptor must be able to react *specifically* with an analyte of interest. For example, a bioreceptor for cholesterol should react only with cholesterol and not with any other compound in the sample. Biological recognition systems such as enzymes or antibodies offer this high specificity and, in addition, ensure high sensitivity and fast response. Usually, the bioreceptor molecules are *immobilised* at, or close to, the surface of the transducer. Immobilisation can be achieved by physical adsorption or entrapment by an inert membrane. The bioreceptor can also be covalently bound to functional groups on the surface of the transducer.

Enzymes are the biomolecules most commonly used as receptors in biosensors. As described in section 5.1.6, enzymes are protein molecules that catalyse chemical reactions. An enzyme can catalyse the conversion of a substrate *A* to a product *B* and itself remain unchanged after the reaction. Hence only *small quantities* of enzyme molecules are required on the surface of a biosensor, as they can regenerate themselves after reacting. Any given enzyme will always turn *A* into *B* and never into *C* and, equally, the same enzyme is extremely unlikely to take *D* and synthesise B. Enzymes are highly *specific* in their action and this *specificity* of enzymatic actions forms the basis for the specificity of the biosensor. Using the enzyme urease results in a sensor for urea and for urea alone, the enzyme will not recognise any other compounds, even if they are chemically related.

Examples of enzymes used for biosensors include glucose oxidase for glucose sensors, alcohol oxidase for ethanol sensors, lactate oxidase for lactate sensors and urease for urea sensors. A typical enzyme reaction, as described by equation 5.2 might involve the transfer of an electron, a pH change, hydrolysis, esterification or bond cleavage. The type of enzymatic reaction that occurs determines the type of transducer that is used.

Antibodies are another type of biomolecule commonly used as bioreceptors. They are also proteins and, as noted in section 5.1.1, they are produced by the immune system of higher animals in response to the entry of foreign materials into the body. Hence, such biosensors are also referred to as *immunosensors*. Antibodies neither catalyse chemical reactions nor do they undergo chemical transformation. They merely undergo a physical transformation by binding tightly to the foreign material, the antigen. Antibodies are very *specific* in recognising and binding only to foreign substances and not to materials that are native to the organism. The antibodies used in biosensors are chosen specifically to target the substance of interest. For example, in a cortisol sensor, the antibody anti-cortisol is used. The sensor can be exposed to untreated blood and, if cortisol is present, it will bind to the anti-cortisol on the sensor-surface. This binding event can be picked up by the transducer e.g. by sensing changes in mass or optical parameters.

Enzymatic bioreceptors have an advantage insofar as the enzyme regenerates itself after reaction. The enzyme is then available for further reaction with the sample. Thus, the response output is directly related to concentration changes in the sample. Antibodies on the other hand can only be used for a one-time measurement. They have to be disposed after reaction or the antigens have to be washed off with suitable reagents.

The bioreceptor does not have to be an enzyme or antibody, virtually any compound that exhibits molecular recognition for an analyte is suitable. This could be a piece of *DNA*, a *cell*, a *microorganism*, an *organelle* or a plant or even mammalian *tissue*. Enzymes and antibodies are used most often, as they are relatively simple to incorporate into a device. This is more difficult with tissue slices and biological cells as they must be supplied with nutrients and have waste fluids removed in order to keep them alive.

5.2.2 *Transducers*

The transducer is essentially the *detector* of the biosensor. It is the component that responds to molecular recognition and converts this response to an output that can be amplified, stored or displayed.

Table 5.3. Forms of transducers.

Electrochemical	electron transfer reactions
amperometric	detects a change in current at a constant potential, for example, O_2/H_2O_2 generated by enzyme reactions
conductometric	detects a change in conductivity between two electrodes, pH or pIon measurement
potentiometric	detects a change in potential at a constant current
Optical	detection of sample spectra or light scattering using optical fibres
Mass	piezoelectric, a quartz crystal changes frequency in response to minor changes at the surface, such as an antibody binding to an antigen
Temperature	exothermic or endothermic reactions

The type of molecular recognition reaction determines the form of the transducer used (Table 5.3). Enzymatic reactions often involve an electron transfer. This electrical activity can be measured with *amperometric, potentiometric* or *conductometric* sensors. If the bioreaction includes the generation of H^+ or OH^- ions, then a pH sensitive dye in combination with an *optical* device can be used. For antibody–antigen binding, the mass change on the surface of the transducer can be detected with a *piezoelectric* device. Exothermic or endothermic reactions can be followed with a *temperature* sensor.

The transducer can convert the signal from the recognition event to an output signal either directly or through a chemical *mediator*. This will be explained using a glucose sensor as an example in the following section (section 5.2.3).

The functioning of a biosensor can thus be summarised as shown in Fig. 5.18. The analyte is recognised by the bioreceptor, which is usually a protein such as an enzyme or antibody. The protein is in close proximity to the detector. This transduces the recognition event into a signal, which can be amplified and displayed.

5.2.3 *The Blood Glucose Sensor*

The blood glucose sensor is the most successful commercial biosensor developed so far. It is used for home testing by individuals suffering from diabetes. About 5% of the population in western countries suffer from this condition and most of them are required to control their blood glucose level several times a day. Thus, a

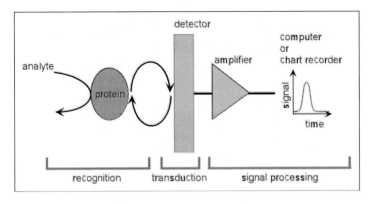

Fig. 5.18. Operating principle of a biosensor.

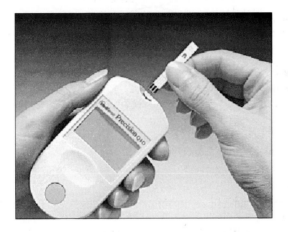

Fig. 5.19. Photograph of an amperometric glucose biosensor (courtesy of Medisense).

rapid, accurate, compact and user-friendly sensor has an immense market potential. An example of such a handheld *amperometric glucose biosensor* is shown in Fig. 5.19.

The highly specific enzyme *glucose oxidase (GOx)* is employed as a bioreceptor. Glucose reacts with this enzyme and generates a redox-acitive species, which can be measured electrochemically (Fig. 5.20). In the first reaction step, glucose reacts with the oxidised form of GOx. Glucose is converted to gluconic acid whilst GOx is reduced. This reduced form of the enzyme then reacts with the oxidised form of a *chemical mediator* to regenerate GOx for further reaction with glucose. The reduced form of the mediator is then oxidised at the electrode

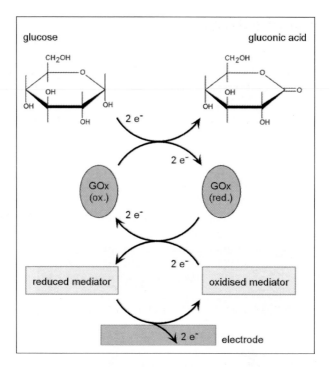

Fig. 5.20. Operating principle of the amperometric detection of glucose.

giving a current signal and regenerating the mediator for further reaction with reduced GOx.

Typically, ferrocenium/ferrocene are used as oxidised/reduced mediator couple. In the absence of the mediator, oxygen, O_2, and GOx (red.) react to form hydrogen peroxide, H_2O_2, and GOx (ox). The mediator ferrocene can be re-oxidised at the electrode at much less extreme potentials than H_2O_2 and thus background current from other blood components can be minimised.

GOx has proven to be an almost ideal bioreceptor. It can be produced cheaply by soil fungi and it withstands greater extremes of pH, ionic strength and temperature than many other enzymes. Also it reacts readily over the concentration range of glucose encountered in human blood samples. Furthermore, the oxidation current is directly proportional to the amount of glucose in the sample.

Opperating a sensor of the type shown in Fig. 5.19 is straightforward. A drop of blood is applied to a disposable electrode strip, which can be inserted into the device. The electrical current is read after 20 s and this is converted to a glucose concentration, which is then displayed on the instrument.

5.3   DNA Binding Arrays

The interaction of DNA, RNA and proteins in organisms is of enormous complexity. Understanding this system is a challenge to be met by scientists and it requires analytical tools for screening and sequencing DNA and RNA. In traditional biomolecular methods one gene is analysed at a time. This becomes very time consuming and tedious when looking at differences between large populations of organisms. In recent years a new technology has become available which allows massively parallel analysis on a single device, the so-called *DNA chip* or *DNA microarray*.

This technique makes use of the molecular recognition of two strands of oligonucleotides which only bind to each other (hybridise) if they are complementary to each other (Fig. 5.21). With a DNA chip it is possible to *identify the sequence* of a gene and discover gene mutations or so-called single nucleotide polymorphisms (SNPs); the inter-individual differences in the genome. Also, a DNA chip can be employed to determine the *expression level* of RNA and the abundance of genes that cause this expression of RNA. Such information is useful for disease diagnosis, drug discovery studies and toxicological research. On one single DNA microarray, tens of thousands of reactions can be performed simultaneously.

5.3.1   *The Principle of DNA Arrays*

In general, an array is an orderly and systematic arrangement of samples. In the case of a DNA array, these are large numbers of DNA molecules or oligonucleotides which are immobilised onto a substrate like a glass slide or a nylon membrane in the form of spots.

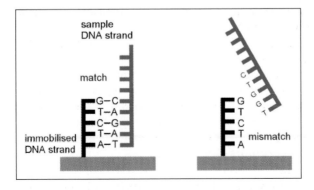

Fig. 5.21. The principle of molecular recognition in DNA array reactions.

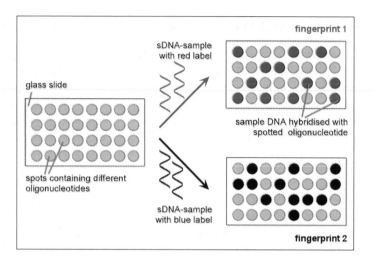

Fig. 5.22. A DNA array is an orderly arrangement of immobilised oligonucleotides on a glass slide, each grey spot represents a different oligonucletide. When reacted with labelled DNA samples, they hybridise with only certain spots on the array, i.e. those containing a matching oligonucleotide sequence. This results in a characteristic pattern, a fingerprint, of coloured and uncoloured spots.

An example of an array is shown in Fig. 5.22. Each grey spot refers to a different type of oligonucleotide immobilised onto the glass substrate. The array is treated with a sample solution containing single stranded DNA fragments with a red label. In the case of a matching sequence, the sample DNA hybridises to the immobilised DNA fragments, as shown in Fig. 5.21. In the event of a mismatch, the label stays in solution and is washed away in the next reaction step. As a result, non-hybridised spots on the array remain colourless while the hybridised ones take on the colour of the label. If another, identical array with the same arrangement of oligonucleotides is treated with a DNA sample containing a blue label then the distribution of coloured and non-coloured spots is different (Fig. 5.22, bottom). The array pattern of labelled and unlabelled spots is characteristic for each DNA sample and is often referred to as its *"fingerprint"*. Once the reaction is completed, the array can be analysed by an imaging software and information about the sequence can be extracted. Fluorescent labels are often used as they can be detected with a high level of sensitivity.

5.3.2 *Fabrication of DNA Arrays*

Arrays come in many different sizes and forms. For standard applications they can be purchased ready to use with the appropriate oligomers already immobilised. For more specialised applications, arrays often have to be custom-made or prepared in-house.

A general differentiation can be made between macro- and microarrays. In *macroarrays*, the spot size is 300 μm or larger. The oligonucleotides to be immobilised are dropped onto the substrate, usually glass, with a small pipette or with a piezoelectric device similar to an ink-jet printer. The array can be read out with fairly conventional scanners. These macroarrays can be home-made on microscope glass slides or custom-made with 1,000 to 10,000 different spots per glass slide. In *microarrays*, spot sizes of 20 to 50 μm are common. Affymetrix Corp. have developed a method to synthesise complete sets of oligonucleotides directly on chip using photolithography, a technology routinely used in the microelectronic industry to make circuit boards, in combination with light activated reactions and combinatorial chemistry. Oligomers containing 25 nucleotides can be readily synthesised. Each spot or *feature* has an area of 20 μm × 20 μm and contains millions of copies of the same oligonucleotide. The whole *GeneChip™* consists of hundreds of thousands of different features (Fig. 5.23). Such small devices require specialised readout equipment, for instance, a confocal laser fluorescence scanner and a dedicated software for analysis.

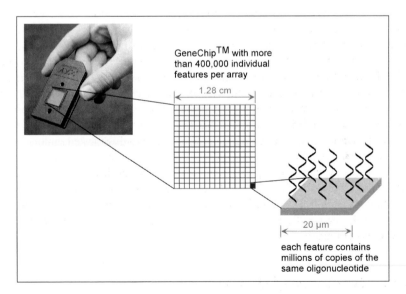

Fig. 5.23. The Affymetrix GeneChip™ consisting of more than 400,000 features, with each feature containing millions of copies of the same oligonucleotide (photo courtesy of Affymetrix Corp.).

5.3.3   Development and Analysis of a DNA Array

The process of carrying out a DNA array can be divided into (1) sample preparation, (2) hybridisation, (3) scanning and (4) data analysis.

DNA samples for genomic studies are extracted directly from cells (section 6.1). For expression studies, RNA is isolated and then converted to cDNA in a process called *reverse transcription* (section 6.2.5). Samples must be fluorescently labelled, denatured to single strands and usually partially digested to shorter DNA fragments.

A few μL of sample solution is either dropped onto the array or passed over it in a flow cell. The hybridisation process takes 12 to 16 hours, usually at elevated temperatures of around 60 °C. After hybridisation, several washing steps must be performed.

The developed array must then be scanned. Macroarrays can be interrogated with a conventional scanner. Microarrays require more dedicated equipment for example laser fluorescence confocal microscopes. The scan can then be analysed with specialised software. Often results are compared using electronic libraries.

5.3.4   DNA Sequencing with Arrays

A good way of understanding the principle of sequencing with a DNA array is to look at the tetranucleotide *CTGA* with its complementary strand *GACT*. An array containing trinucleotides can be prepared to determine this sequence. There are $4^3 = 64$ different trinucleotides, hence, an array with 64 spots is required. The sample DNA is partially digested and labelled with, for example, a fluorescent marker. A signal is obtained if the added DNA hybridises with the DNA on the spot. In the case of the CTGA tetranucleotide, the partially digested sample strand will hybridise with the *GAC* and *ACT* trinucleotides. Due to overlapping, the sample sequence can be unambiguously identified as GACT.

In Figs. 5.24 and 5.25, a slightly more complex example is shown for a longer sequence of sample DNA and an array with tetramers consisting of $4^4 = 128$ spots. Each spot contains multiple copies of the same tetramer. When the array is treated with labelled, partially digested sample DNA, hybridisation occurs in certain spots (Fig. 5.25, right). These spots have to be identified and ordered according to their overlapping bases. From this, the sequence of the sample can be determined (Fig. 5.25, left).

Constructing arrays with longer oligomers enables the analysis of longer stands of sample DNA. However, fabrication of the array becomes more complex as the required number of spots increases exponentially. For example, octanucleotides

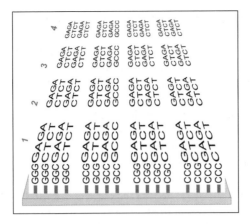

Fig. 5.24. An array consisting of tetranucleotides with $4^4 = 128$ spots. Although only one strand is depicted per spot, each one contains many copies of the same tetranucleotide.

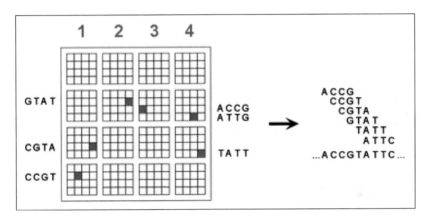

Fig. 5.25. The array as shown in Fig. 5.24 after reaction with a partially digested DNA strand. Hybridisation occurred in six different spots, highlighted in red. After identifying the base sequences of these spots and sorting them, the sequence of the sample DNA strand can be reconstructed.

require an array with $4^8 = 65,536$ different spots. Such a high number of spots over a very small area emphasises the need for robotics and micromachining techniques for array fabrication. With octanucleotides, sequences of about 200 base pairs can be readily determined.

5.3.5   *Other Applications of DNA Arrays*

Arrays can generally be designed for two different modes: (1) the *immobilised* nucleic acid is of a *known sequence* as in the examples shown above and reacted with an unknown labelled sample sequence or (2) the *unknown* sample sequence is *immobilised* and is reacted with labelled oligomers. Mode (1) is best suited for sequencing, as here it is necessary to react the sample with a large number of oligomers with different sequences. Method (2) is better for analysing a large number of different samples using the same reagent.

The immobilised molecule is either an *oligonucleotide* or a strand of *cDNA* (section 6.2.5). Oligonucleotides consist typically of between 25 and 70 nucleotides. These can either be prefabricated and spotted onto the array surface or directly synthesised on the glass surface, as outlined for the Affymetrix GeneChip™. cDNA is obtained from isolated RNA by reverse transcription (section 6.2.5). After PCR amplification, cDNA strands of between 500 and 5000 bases are usually immobilised on the array substrate.

In addition to DNA sequencing as described above, DNA arrays have a number of other applications in diagnostics, pharmacogenomics, expression profiling and toxicology to name a few. Normal cells can be compared to diseased cells or cells treated with drugs. DNA chips allow comparison of genomic DNA of different cells for gene discovery and polymorphism analysis or comparison of RNAs for expression profiling.

The patented Affymetrix GeneChip™ is a very powerful and reliable platform, allowing massively parallel analysis. However, this comes at a cost, that is frequently beyond the budget of most academic research laboratories. Home-made arrays have a much lower density of spots and also tend to be less reliable when compared to the stringent controls applied by Affymetrix. They are, however, much cheaper and can be designed specifically for each particular application. As competing companies try to fill the gap for more affordable array platforms, prices will fall and DNA arrays will find increasing use both in academic and in industrial laboratories.

5.4   DNA Identification by Pyrosequencing

Pyrosequencing was developed in the late 1990s by Mostafa Ronaghi at the Royal Institute of Technology in Sweden. It is already used in research laboratories worldwide. The method is based on a *chemiluminescent enzymatic reaction*, which is triggered when a molecular recognition event occurs. Essentially, the method allows sequencing of a single DNA strand by *synthesising the complementary strand* along it. Each time a nucleotide, A, C, G or T is incorporated into the growing chain a cascade of enzymatic reactions is triggered which results in a light signal. DNA stretches up to 50 or 60 bases can be determined by pyrosequencing.

50 bases might not sound very significant in comparison to conventional sequencing methods (see chapter 6), especially when bearing in mind that there are billions of bases in the human genome. However, *short stretches* of the DNA sequence are frequently sufficient to characterise a particular virus or bacteria or to analyse differences between healthy and diseased individuals. Since pyrosequencing is fully automated, fast, reliable and accurate, large numbers of samples can be analysed in a short time. In a single day, this could be as high as tens of thousands of samples.

5.4.1 *The Principle of Pyrosequencing*

The purpose of pyrosequencing is to determine the base sequence of a short DNA strand by synthesising the complementary strand along the template. This process can be divided into several steps.

First, a single stranded DNA template, usually PCR amplified and purified by chromatography, gel filtration or electrophoretic techniques, is *immobilised* onto a surface. A suitable *primer* is then hybridised to this single strand (Fig. 5.26). This ensemble is incubated with four enzymes (DNA polymerase, ATP sulfyrase, luciferase and apyrase) and two substrates (adenosine $5'$ phosphosulfate (APS) and luciferin).

Next, one of the four deoxynucleotide triphosphates (dNTPs), dATP, dGTP, dCTP or dTTP, is added to the reaction mixture. If the nucleotide is complementary to the next base in the strand, then DNA polymerase catalyses its incorporation. This reaction is accompanied by the release of a pyrophosphate (PPi) molecule (Fig. 5.27). The amount of PPi released is equimolar to the amount of nucleotide incorporated. For example, if three nucleotides are incorporated, then three molecules of PPi are released.

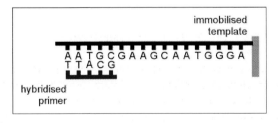

Fig. 5.26. A single stranded DNA template is immoblised onto a surface and hybridised with a suitable primer.

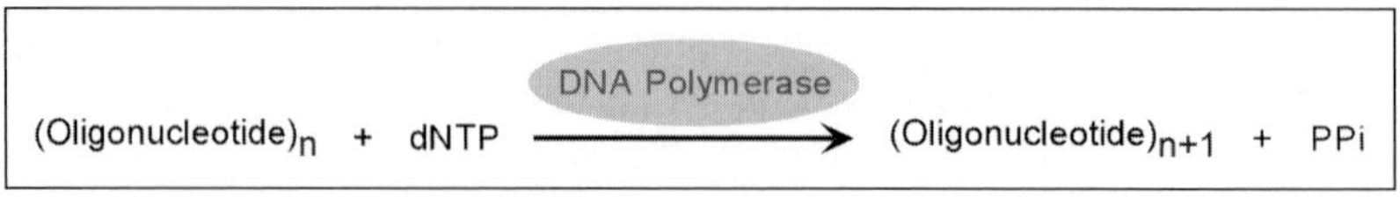

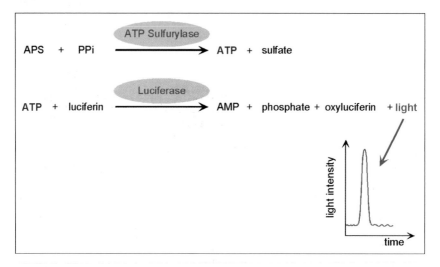

Fig. 5.27. Incorporation of a deoxynucletide triphosphate (dNTP) into the oligonucleotide or DNA chain is catalysed by DNA polymerase and results in the release of
pyrophosphate (PPi).

Fig. 5.28. The cascade of enzymatic reactions resulting in the chemiluminescent release
of light, which can be detected with a CCD camera and is plotted in the form of a
pyrogram™.

PPi then triggers a cascade of enzyme-catalysed reactions (Fig. 5.28). The
substrate APS reacts with the PPi and is catalysed by ATP sulfurylase, to form
adenosine triphosphate (ATP). The enzyme luciferase catalyses the reaction of
ATP with the substrate luciferin, generating oxyluciferin and visible *light*. This
light can be detected by a charged coupled device (CCD) camera as a peak in a
so-called *pyrogram™*. The signal obtained is directly proportional to the amount
of dNTP incorporated.

Any excess nucleotide dNTP and any excess ATP are degraded by the nucleotide
degrading enzyme *apyrase* to their respective mono-and diphosphates (Fig. 5.29).
When degradation is complete, the next dNTP can be added.

The addition of dNTPs is performed iteratively, one at a time. The next dNTP
can only be added once all the nucleotide from the previous step has been degraded.
During this process of iterative nucleotide addition, a complementary DNA strand

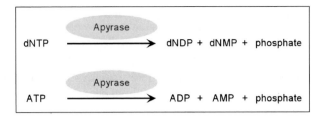

Fig. 5.29. The nucleotide degrading enzyme apyrase destroys any remaining dNTP and ATP.

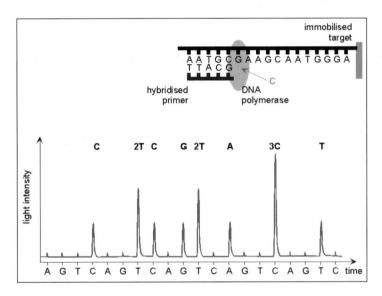

Fig. 5.30. Pyrogram™ obtained from sequencing the immobilised target DNA.

is synthesised, one base at a time and the sequence can be read from the obtained pyrogram™ (Fig. 5.30). It should be noted that the natural dATP was found to interfere with the luciferase reaction and so deoxyadenosine α-thio triphosphate (dATP-αS) is used instead.

A limitation of pyrosequencing is that large numbers of the same base in a row, *homopolymeric regions*, cannot be detected easily. The light signal is, as mentioned above, proportional to the number of nucleotides incorporated. However, this is only true for up to 5 or 6 bases. Stretches of up to 10 of the same base in a row can still be analysed using dedicated software. Homopolymeric regions longer than 10 bases cannot be resolved.

5.4.2   *Sample Preparation and Instrumentation*

The DNA of interest must be isolated from the cell and amplified using PCR (section 6.2). Purification must be performed, for example via chromatography, to remove primers, unincorporated nucleotides and salts remaining from the PCR procedure. The purified DNA is then immobilised onto a solid surface, normally onto streptavidin-coated microspheres. The streptavidin binds to biotin labels that have been attached to one strand of the double stranded DNA. The unlabelled strand is denatured leaving the beads coated with single stranded DNA ready for pyrosequencing.

The immobilised sample DNA is filled into the 96 or 384 wells of a microtiter plate-style platform. This platform is mounted onto a stage which is temperature controlled and slightly agitated for improved sample mixing. Small volumes of reagents and nucleotides are delivered by a disposable ink-jet type cartridge.

A lens array is used to focus light from each well to a specific area on the CCD camera for recording of the pyrograms. Thus 96 or even 384 sequencing reactions can be performed in parallel in a fully automated fashion in less than an hour.

5.4.3   *Applications of Pyrosequencing*

Pyrosequencing allows the determination of short stretches of DNA up to 50 or 60 base pairs. The list of applications for this type of sequencing is long and includes (1) typing of SNPs, (2) typing of viruses, bacteria and fungi, (3) sequencing of disease associated genes.

Determination of *SNPs* is likely to become the main application of pyrosequencing. SNPs are variations in the DNA sequence from one individual to the next. Most commonly, a single nucleotide in the genome is altered, for example instead of the sequence GCTAC GGTCAG the altered sequence GCTA T GGTCAG is found. An SNP is found in every few hundred bases. Some SNPs occur in noncoding areas of the DNA and are irrelevant for biochemical processes. Others occur in coding areas and can be responsible for disorders or disease susceptibility. The human genome project has recently decoded the sequence of human DNA. Now it is necessary to identify the coding parts of the DNA (97 % of human DNA is estimated to be non-coding), to identify differences between individuals (99 % of human DNA is identical from one individual to the next) and to discover their relevance for disease development. More than one million SNPs have already been catalogued. Pyrosequencing is a method that allows automated and parallel processing of a large number of samples making it a popular method for SNP analysis.

Summary

The application of molecular recognition in bioanalytical chemistry has been described in this chapter. The specific binding of antibodies and antigens is used in immunoassays. A large number of assay formats are possible including competitive and non-competitive assays, heterogeneous and homogeneous assays, labelled and un-labelled assays.

Biosensors make use of the biorecognition of antibodies and antigens as well as the high specificity of enzymes. Molecular recognition and signal transduction are combined into a single device. The blood glucose sensor was described as an example.

Hybridisation of matching strands of ssDNA is the underlying principle of DNA arrays. Modern array platforms, such as that developed by Affymetrix Corp., are extremely miniaturised and allow massively parallel analysis for DNA sequencing and comparative studies of DNA samples.

Pyrosequencing uses the specific molecular recognition of base pairs. This is combined with a cascade of enzymatic reactions yielding in a chemiluminescent signal. The method allows fully automated sequencing of short stretches of DNA mainly for SNP analysis.

References

1. P. C. Price and D. J. Newman, *Principles and Practice of Immunoassay*, 2nd edition, Stockton, 1997.
2. A. L. Lehninger, D. L. Nelson and M. M. Cox, *Principles of Biochemistry*, 2nd edition, Worth Publishers, 1993.
3. A. Manz and H. Becker, Eds., *Microsystem Technology in Chemistry and Life Sciences*, Springer-Verlag, 1998.
4. S. B. Primrose and R. M. Twyman, *Principles of Genome Analysis and Genomics*, 3rd edition, Blackwell Publishing, 2003.
5. J. Gosling (editor), *Immunoassays: A Practical Approach*, Oxford University Press, 2000.

Chapter 6

NUCLEIC ACIDS AMPLIFICATION & SEQUENCING

In this chapter, you will learn about. . .

♦ . . .how different nucleic acids can be extracted and isolated.

♦ . . .nucleic acid amplification using the polymerase chain reaction.

♦ . . .reverse transcription of an RNA strand into a strand of complementary DNA (cDNA).

♦ . . .the different techniques for determining the base sequence of nucleic acids.

Nucleic acids, *deoxyribonucleic acid* (*DNA*) and *ribonucleic* (*RNA*), are among the most studied molecules in biochemistry. DNA, and in some cases RNA, is the carrier of genetic information in all living organisms. The study and analysis of the genetic code can provide information that can be used in clinical diagnostics, drug discovery, genetic engineering and forensic sciences, to name a few.

This chapter is divided into three main sections. First, some of the techniques for *extracting* and *isolating* DNA and RNA contained in cells are presented. Next, the *Polymerase Chain Reaction* (PCR); is explained. With this method, DNA molecules can be amplified. A single molecule is sufficient to make millions of copies. RNA cannot be amplified by this method. A sequence of RNA can, however, be "rewritten" as a base sequence of cDNA by a process called reverse transcription. In the third section of this chapter, the different techniques for *sequencing of nucleic acids* are described. The DNA molecules are partially fragmented with a *restriction enzyme*. The sequence of the obtained fragments can then be determined via the *Maxam-Gilbert* method or the *Sanger* method.

6.1   Extraction and Isolation of Nucleic Acids

The most commonly analysed nucleic acids are chromosomal and plasmid DNA as well as messenger RNA (mRNA). Numerous methods have been reported for the extraction and isolation of these nucleic acids. The isolation process usually consists of three steps: (1) *Cell lysis*: To release the contents of the cell, the cellular walls are ruptured for example by treatment with a hypotonic solution or by treatment with surfactants such as sodium deodecyl sulfate (SDS) or Triton X-100. Alternatively, the cells are incubated with an enzyme that has cell lytic properties, such as lysozyme. (2) *Isolation*: The nucleic acids are isolated by removing cell debris, separating the nucleic acids from other cell contents and breaking down any complexes between proteins and nucleic acids. Methods such as centrifugation, precipitation, size exclusion chromatography (SEC, section 2.3.4), liquid-liquid chromatography, treatment with protein digesting enzymes and CsCl density gradient centrifugation are all commonly used. (3) The DNA or RNA of the resulting solution is then *concentrated* or amplified. Some of these isolation and extraction methods are explained in more detail in the following sections.

6.1.1   *CsCl density gradient centrifugation*

Caesium chloride density gradient centrifugation can be used to isolate DNA. Due to the difference in *buoyant densities*, DNA is separated from RNA or proteins. Such a separation can be achieved by centrifuging the crude sample in a tube containing a density gradient formed by a CsCl solution. Upon centrifugation, the different macromolecules in the sample form distinct bands depending on their buoyant densities. In the example shown in Fig. 6.1, the density gradient ranges from 1.6 to 1.8 g mL^{-1}. RNA has a relatively high density and sinks to the bottom of the tube. Proteins have a relatively low density and float on the top. The buoyant

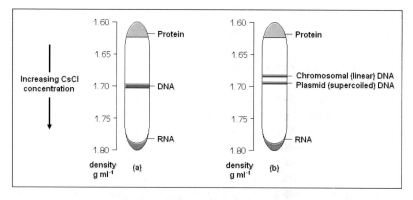

Fig. 6.1. (a) CsCl density gradient centrifugation separates DNA from RNA and proteins due to their different buoyant densities. (b) By addition of ethidium bromide, chromosomal DNA can be separated from plasmid DNA.

density of DNA is about $1.7 \, \text{g} \, \text{mL}^{-1}$. DNA molecules are thus concentrated close to the middle of the tube, where the CsCl density equals $1.7 \, \text{g} \, \text{mL}^{-1}$.

CsCl density gradient centrifugation is also capable of separating chromosomal from plasmid DNA by centrifugation in the presence of ethidium bromide. This fluorescent *intercalator* binds between the two DNA strands (see also section 6.2.4). Upon binding, it causes a decrease in buoyant density due to the partial unwinding of the double helix. More ethidium bromide is bound to the linear, chromosomal DNA than the supercoiled, plasmid DNA. The density of chromosomal DNA decreases by $0.125 \, \text{g} \, \text{mL}^{-1}$, whereas the density of plasmid DNA decreases by only $0.085 \, \text{g} \, \text{mL}^{-1}$. This difference is sufficient to resolve the two types of DNA. Ethidium bromide fluoresces upon irradiation with UV light. This effect is used to visualise the two bands. The isolated DNA can be removed from the centrifugation tube with a syringe.

6.1.2 *Total Cellular DNA Isolation*

For isolating all DNA contained in a cell, the cell culture or tissue sample is transferred into a buffer which contains a detergent such as SDS or Triton *X*-100. The detergent disrupts the cellular walls and dissociates any DNA-protein complexes. RNA molecules contained in the cell extract are broken up by treatment with a ribonuclease. Proteins can be digested by a proteolytic enzyme, most commonly proteinase *K*. The DNA can then be extracted from the mixture by precipitation with ethanol. Only long nucleic acid chains precipitate, single nucleotides and products of the RNA digestion remain in solution.

6.1.3 *RNA Isolation – The Proteinase K Method*

The isolation of RNA is more challenging than that of DNA. Samples can be easily contaminated by the ubiquitous ribonuclease, which causes RNA to be fragmented. Hence, the use of sterile glassware and chemicals of high purity is imperative. Additionally, RNA forms tight complexes with proteins. Coarse treatment is required to release RNA from these complexes.

A common method for RNA isolation is the *proteinase K method*. In this method, the cells are lysed by incubation in a hypotonic solution followed by centrifugation to remove DNA and cell debris. Treatment with the proteolytic enzyme proteinase *K* leads to the dissociation of RNA-protein complexes and the digestion of the proteins. The digestion products are then removed by phenol/chloroform extraction and the RNA in the remaining aqueous solution is precipitated using ethanol.

6.2 Nucleic Acid Amplification – The Polymerase Chain Reaction (PCR)

The *Polymerase Chain Reaction (PCR)* is an *in vitro* technique that allows the enzymatic-catalysed amplification of specific nucleic acid sequences. The method was conceived by Kary Mullis in 1983, allegedly at the side of a California mountain road while he was resting. It was presented for the first time in 1986 at the 51st Cold Spring Harbour Laboratory Symposium on Quantitative Biology. In 1993, Mullis was awarded the Nobel Prize for his invention.

PCR is an immensely powerful tool that has revolutionised modern biochemistry and has had a tremendous impact on genetics and medical diagnostics. With PCR, minute quantities of nucleic acids can be analysed. Even a single DNA molecule is enough for amplification. Often, the available amount of a DNA sample is very small and not sufficient for any sequencing and analysis. With PCR amplification, many formerly unsolvable tasks are now routine. In forensics, a single hair or sperm is enough to identify a donor. In clinical medicine, infectious diseases can be identified rapidly. Variations and mutations in the genetic code of individuals within a species or between different populations of a species can be studied relatively easily.

The *principle of PCR*, including the different phases of the amplification reaction and the reagents needed, is described in the next section. This is followed by *real time PCR*, which allows semi-quantitative measurement of the PCR products during the reaction. Then the process of *reverse transcription* of RNA into cDNA is outlined.

6.2.1 The Principle of PCR

The amplification reaction is carried out in a single sample tube, which is placed into a thermocycler. The sample tube contains (1) an excess of two *primers*, i.e. oligonucleotides, that are complementary to the ends of the targeted nucleic acid region, (2) the enzyme *DNA polymerase*, (3) an excess of the four *deoxynucleotide triphosphates* (dATP, dCTP, dGTP and dTTP) (4) the template DNA, i.e. the DNA sample that is to be amplified and (5) a buffer to maintain the correct pH and to supply ions such as Mg^{2+} their are necessary for the reaction. The reaction takes place in three temperature-controlled steps (Fig. 6.2).

- *Step 1 – DNA denaturation:* The temperature of the reaction mixture is elevated to 94–95 °C. The double stranded sample DNA (dsDNA) is denatured into single strands (ssDNA), as the hydrogen bonds between the complementary bases are broken. The polymerase enzyme is inactive at this temperature.
- *Step 2 – Primer annealing:* The temperature is then lowered to between 50 and 60 °C. H-bonds can be reformed at this temperature and the single DNA

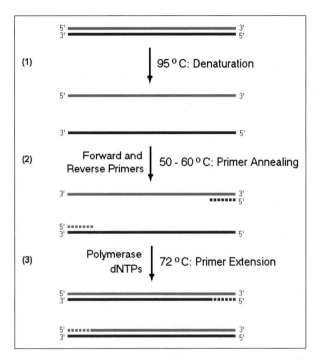

Fig. 6.2. The principle of PCR. (1) Denaturation: the two DNA strands are separated at 95 °C; (2) Annealing of primers: the primers are hybridised to their complementary sequences at 50–60 °C; (3) Primer extension: at 72 °C, the polymerase catalyses the synthesis of the complementary single stranded DNA by extending the 3′-end of the hybridised primer.

strands could hybridise with their complementary strands to reform dsDNA. This, however, is highly unlikely due to the vast excess of primer molecules, $10^{10} - 10^{11}$ primers to a few DNA template molecules. Hence, it is the primers that anneal with the complementary region on the target DNA. The primers consist of a sequence of about 10–30 bases which is specifically targeted towards the sample DNA. The primers are chosen such that they hybridise with the 3′-ends of a sample DNA.

- *Step 3 – Primer extension:* The synthesis of DNA is catalysed by the polymerase. The 3′-end of the hybridised primer is extended along the original DNA strand by continuous incorporation of the complementary nucleotides into the chain. Thus, a complementary DNA strand is formed. Primer extension proceeds from the 3′-end to the 5′-end. This extension reaction stops as soon as the complementary strand has been completed. It has been empirically shown that the fidelity of the construction of the complementary strand improves at 72 °C. Hence, the temperature is usually adjusted to this value for the primer extension step.

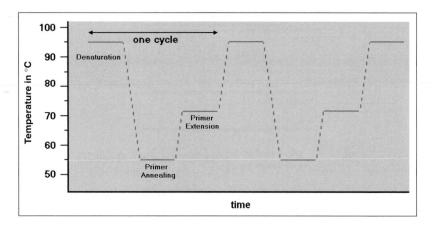

Fig. 6.3. The temperature profile of the reaction. A typical PCR consists of 20 to 35 cycles.

These three stages constitute *"one cycle"*. A typical PCR consists of 20 to 35 such cycles. This procedure of repeated heating and cooling of the reaction mixture is commonly referred to as *thermal cycling* (Fig. 6.3). The denaturation, annealing and extension steps are carried out for about 30–120 s each. Only the very first denaturation step takes about 5 min. A PCR with 30 cycles can, thus, be completed within 1–2 h. Efficient heating and cooling as well as precise temperature control are required. This can be achieved by Pelletier elements, which operate upon resistive heating and semiconductor cooling. Alternatively, heating and/or cooling can be controlled with air, fluid or irradiation.

In theory, the number of DNA copies is doubled during each cycle, resulting in an exponential amplification. The theoretically predicted number of DNA molecules, N_m, at the end of the reaction depends on the initial number of DNA copies, N_0, in the reaction mixture and the number of cycles, n:

$$N_m = N_0 \cdot 2^n \qquad \text{(equation 6.1)}$$

However, this theoretical number is never achieved due to a number of limiting factors including the depletion of the reagents and the amplification of longer strands during the first cycles (see section 6.2.2). In practice, the number of DNA molecules can be approximated by the following equation:

$$N_m = N_0 \cdot (1 + x)^n \qquad \text{(equation 6.2)}$$

where x is the *efficiency* of the reaction. It can have a value between 0 and 1 or it can be expressed as a percentage.

For a reaction with $n = 20$ cycles that starts with a single DNA molecule, $N_0 = 1$, the number of theoretically predicted copies after the end of the reaction is more than 1 million. Assuming an efficiency of $x = 0.7$, the reaction "only" gives a 40,000-fold amplification.

The efficiency and the yield are not the only measures of the efficacy of PCR. The *specificity* – a highly specific reaction generates only the target sequence – and fidelity – high *fidelity* means that there is a negligible number of polymerase-induced errors in the product – are equally important characteristics of a reaction. Ideally, a PCR has high efficiency, yield, specificity and fidelity.

6.2.2 *The Rate of Amplification During a PCR*

The sample DNA molecules are usually isolated and extracted from cells as outlined in section 6.1. In most cases, it is not necessary to amplify the whole DNA molecule; a section of the DNA strand is generally sufficient. The primers are chosen such that they hybridise at each end of this targeted section of interest. The amplification of target length DNA molecules proceeds in three phases: (1) a starting phase in which no target length molecules are produced, (2) a phase of exponential growth and (3) a plateau phase.

The first four cycles of a PCR are depicted in Fig. 6.4. During the first annealing step, primers hybridise at one end of the anticipated target section. The complementary DNA strand is synthesised from this point to the end of the original DNA strand. This cDNA strand is, thus, shorter than the original strand but still longer than the anticipated target region. Hence, during the first cycle, no target copy is produced. During the second cycle, primers anneal at the other end of the target region. Extension results in the first target length single DNA strands. At the same time, more non-target length molecules are produced.

Two molecules of dsDNA with target length are synthesised only during the third PCR cycle. The presence of these target length molecules at the beginning of cycle four sparks the *geometric (exponential) amplification*. At the end of cycle four, there are already eight target length dsDNA molecules. At the end of cycle five, this number increases to 32. During every cycle, molecules are produced, that are longer than the target region. However, their amplification rate is linear and, thus, much slower than that of the target length DNA molecules. The amount of these by-products at the end of the reaction is negligible compared to the amount of target molecules.

As the reaction proceeds, a *plateau phase* is reached. Continued thermal cycling does not lead to the production of significant amounts of product anymore. One reason for this is the depletion of the reagents, i.e. the depletion of primers and nucleotides. Furthermore, phosphates, that are released during the reaction act as inhitors to the enzyme. Also, the DNA polymerase deteriorates after repeated cycling. Another limiting factor is that hybridisation of longer strands of DNA occurs at higher temperatures than hybridisation of a ssDNA and a primer. Thus,

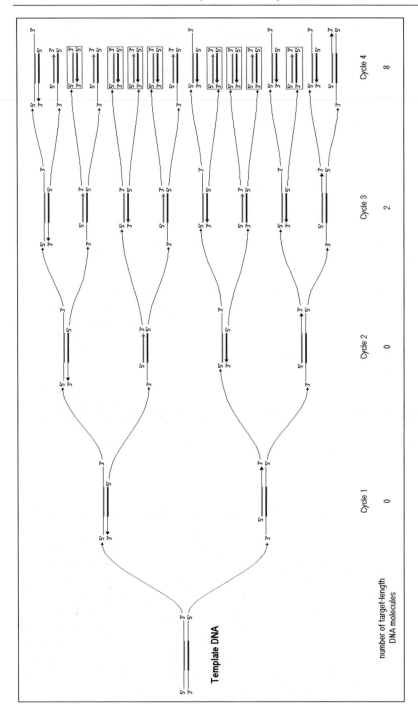

Fig. 6.4. Amplification during the first four cycles. The first target-length molecules appear after the third cycle.

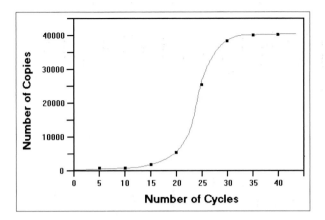

Fig. 6.5. The amplification profile of a PCR showing the slow start, exponential amplification and plateau phase.

as the reaction mixture is cooled down after denaturation from 95 to 60 °C, two ssDNA can hybridise to form dsDNA before any primer annealing can take place.

The amplification rate during these three phases is shown in Fig. 6.5.

6.2.3   *Reagents for PCR*

The reagents required for PCR, apart from the DNA template itself, are the DNA polymerase enzyme, an appropriate pair of primers and the four nucleotides (dNTPs). For the PCR to be successful, the reaction conditions have to be carefully controlled, including pH, ionic strength and additives. The reagents used for PCR are outlined in more detail below.

6.2.3.1   *DNA Polymerase*

The role of the polymerase enzyme is to catalyse the synthesis of complementary DNA strands. When PCR was first developed, the enzymes used were heat-labile and degraded at the elevated temperatures during denaturation. Hence, fresh enzyme had to be added during each cycle. The reaction was facilitated by the introduction of heat-stable polymerases, such as *Taq polymerase*, which originates from the bacterium *Thermus aquaticus*. Other modern polymerases include Tth-, Pwo and Pfu polymerase. Apart from being heat-stable, the enzyme should also be capable of synthesising long stretches of DNA, i.e. have a good *polymerisation activity*. Pwo and Pfu Polymerase additionally exhibit a proof reading activity, called $5'-3'$ *exonuclease activity*. Incorrect nucleotides are recognised, removed and replaced by the correct ones. The $5'-3'$ exonuclease activity, thus, improves the fidelity of the replication.

6.2.3.2   *Primers*

Primers are short oligonucleotides, which are complementary to the ends of the target sequence. Two distinctive primers are used for a PCR amplification: a forward primer and a reverse primer (Fig. 6.2). Each hybridises to one of the two strands of the original dsDNA molecule (see Fig. 6.4, cycle one). DNA synthesis always proceeds in the direction from 5' to 3'. The first nucleotide to be incorporated reacts with the free 3'-hydroxyl group of the primer. The 5'-end of the primer is blocked.

The set of primers has to be designed specifically for the sample DNA. The primer length is usually 10–30 base pairs (bp) and their complementary sequence must be unique in the template. Additionally, there should be no *intra* or *inter* primer complementarity in order to avoid the formation of primer-dimers. Ideally, the number of each base in the primer is relatively equal. Unusual sequences such as long stretches of polypurines or polypyrimidines and repetitive sequences must be avoided. The melting temperature for both primers should be similar and lie between 55 and 80 °C.

6.2.3.3   *Deoxynucleotide triphosphates (dNTPs)*

The four deoxynucleotide triphosphates, dATP, dCTP, dGTP and dTTP, are the building blocks for DNA synthesis (see section 1.2.1.1). The reaction mixture must contain an excess of these dNTPs as they deplete during PCR. The concentration of each dNTP should be equal.

6.2.3.4   *Buffer*

The buffer pH and ionic strength is chosen according to the polymerase used. The ionic strength has a crucial influence on the specificity of the PCR. A typical buffer system has an ionic strength of about 50 mM and consists of Tris-HCl, pH 8.3, with KCl or NaCl.

Magnesium chloride, $MgCl_2$, at concentrations between 0.5 and 5 mM, is always added. The Mg^{2+} ions form a soluble complex with DNA and polymerase. Their role is to bring the polymerase and DNA into close proximity and to balance the negative charges on the DNA molecule. Additionally, they stimulate the polymerase activity. The concentration of the Mg^{2+} ions is related to both specificity and yield of the reaction. At low Mg^{2+} concentrations, the enzymatic activity of the polymerase is decreased. Excess Mg^{2+} results in poor denaturation because dsDNA molecules are stabilised by the Mg^{2+} ions. Furthermore, high magnesium concentrations lead to increased annealing of the primers to incorrect sites and, hence, loss of specificity.

A number of *additives* can be employed to stabilise the polymerase enzyme or to optimise primer annealing. Examples include glycerine, bovine serum albumin

(BSA) and polyethylene glycol (PEG). Denaturation can be improved by adding dimethyl sulfoxide (DMSO), formamide or a surfactant such as Tween-20®. The choice of additives depends heavily on the polymerase used and on the particular DNA template.

6.2.4 Real-Time PCR

Real-time PCR is also referred to as quantitative PCR (QT-PCR). With this technique, the formation of the reaction products can be monitored as the reaction proceeds. Usually a fluorescent marker is employed. The increase in products after each cycle can be recorded as an increase in fluorescence. Such an amplification plot provides a more complete picture of the PCR process than measuring the product accumulation at the end of a fixed number of cycles. Two different types of assays can be used for real-time PCR: *dsDNA binding dye assays* and *probe based assays*.

6.2.4.1 dsDNA binding dye assays

The binding dye is a small molecule that fluoresces upon binding to double-stranded DNA. As PCR amplification produces more and more dsDNA molecules, the fluorescence signal increases (Fig. 6.6). Dyes are classified into two categories

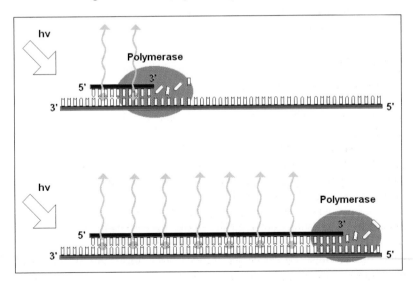

Fig. 6.6. Dye-based real-time PCR with an intercalator: the fluorescence intensity of the intercalator is proportional to the mass of the dsDNA present. As more and more DNA is produced, the fluorescence signal increases.

Table 6.1. Some common intercalators and groove binders.

Intercalators	Groove binders
Ethidium bromide	Distamycin
Daunomycin	Netrospin
Actinomycin D	4,6-diamidino-2-phenylindole
SYBR-green	

(1) *intercalators*, which bind between the two strands of dsDNA, and (2) *minor groove binders*, which bind externally to dsDNA (Table 6.1). For a dye to be suitable, it must exhibit increased fluorescence when bound to dsDNA and it must not inhibit PCR.

As multiple dye molecules bind to one DNA molecule, the method is very sensitive especially during the first cycles. The dyes bind to any dsDNA present. This makes the method very versatile as the same dye may be used for any sequence amplification. On the other hand, the dyes do not distinguish between specific and non-specific dsDNA. Mismatches and primer-dimers also give fluorescence signals.

6.2.4.2   *Probe-based assays*

The probe-based method relies on the degradation of a target probe by the $5'–3'$ exonuclease activity (see section 6.2.3.1) of the polymerase. The probe is a short oligonucleotide, which is complementary to a region of the target sequence between the two primers. A fluorescent molecule, *the reporter* is attached to the $5'$-end of the probe and a *quencher* to the $3'$-end. The proximity of the quencher reduces the fluorescence emitted by the reporter due to *Fluorescence Resonance Energy Transfer (FRET)*. The probe is added to the PCR mixture and hybridises to the ssDNA after denaturation. The fluorescence signal is quenched at this stage due to the proximity of reporter and quencher. During primer extension, the reporter is cleaved from the probe due to the exonuclease activity of the polymerase. Now, no longer in the proximity of the quencher, the reporter fluoresces in the reaction mixture (Fig. 6.7). The intensity of the fluorescence is directly proportional to the number of times this process has been repeated i.e. to the number of amplified molecules.

With the probe based method, only specific products are detected. Multiple probes with different reporters can be used for simultaneous detection of a number of distinct sequences.

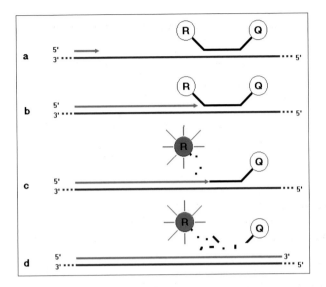

Fig. 6.7. Probe based real-time PCR: (a) During primer annealing the probe hybridises to a region within the target sequence and fluorescence is quenched. (b) and (c) As the primer extension proceeds, the exonuclease activity of the polymerase cleaves the reporter from the rest of the probe. (d) The reporter now fluoresces freely as it is in solution and no longer in proximity of the quencher.

6.2.5   *Reverse Transcription – PCR (RT-PCR)*

Often, RNA is available for analysis, rather than DNA. This is the case in cultures of some viruses e.g. the human immunodeficiency virus (HIV), which only contain RNA. PCR, however, is not capable of using RNA as a template and, therefore, amplification does not take place. In order to analyse genetic material of such species, PCR is combined with another enzymatic reaction called *reverse transcription (RT)*. In RT, the enzyme *RNA-directed DNA polymerase* also known as *reverse transcriptase*, is used. This enzyme synthesises, or more accurately transcribes, mRNA to its complementary strand of DNA (cDNA).

The reverse transcription can be performed in the same reaction vessel as the subsequent PCR. Initially, the reaction mixture contains the RNA template, the reverse transcriptase enzyme as well as a primer directed against the targeted RNA region and the four dNTPs.

The RNA template is denatured at 72 °C and then cooled to 42 °C to allow the primers to anneal. Catalysed by the reverse transcriptase enzyme, the primers are extended in the 5′ to 3′ direction resulting in a strand of cDNA. The reaction mixture is then heated to 94 °C. At this temperature, the reverse transcriptase is inactivated. Finally, the PCR reagents are added for cDNA amplification as described in the previous sections.

6.3   Nucleic Acid Sequencing

As the function of a nucleic acid is determined by the sequence of the bases within the molecule, sequencing plays an important role in nucleic acid analysis. Walter Gilbert and Frederic Sanger are considered as the pioneers of DNA sequencing. In 1980, they were jointly awarded the Nobel prize. For Sanger, this was the second time (see chapter 7). The development of fast and automated sequencing techniques has received considerable interest in the past decade due to the Human Genome Project (HGP). DNA sequencing today is one of the fastest growing technologies in biochemical analysis.

In this chapter, the *Maxam–Gilbert method* and the *Sanger method* for *ab initio* sequencing of long nucleic acid chains are discussed. Two alternative methods for sequencing of shorter DNA strands are described in chapter 5: with DNA arrays, sequences of up to a few hundred base pairs can be determined (section 5.3), while pyrosequencing is capable of identifying sequences of up to 50 bp (section 5.4).

6.3.1   *The Use of Restriction Enzymes in Sequencing*

DNA extracts from cells or tissues are generally too long to be sequenced directly. They must be broken down in an orderly manner, into smaller fragments of up to 800 base pairs. For this purpose, *restriction endonucleases*, or simply *restriction enzymes*, are employed. These enzymes recognise a specific base sequence of four to eight bases within a dsDNA molecule and cleave both strands at a specific point close to this recognition site. Some commonly used restriction enzymes and their recognition sequence and cleavage point are listed in Table 6.2.

Digestion with restriction enzymes produces a series of precisely defined fragments, which can be separated according to their size by gel electrophoresis (chapter 3.2). The fragments then need to be denatured into single strands. This can be achieved by melting the dsDNA. The single strands are then separated by gel electrophoresis. If the DNA fragments are still longer than 800 base pairs after the initial digestion, a further digestion step with a second restriction enzyme is required. A particular DNA molecule can be treated consecutively with several different restriction enzymes or even with a mixture of restriction enzymes. After gel electrophoresis, the lengths of the obtained fragments can then be determined by comparing them to a ladder (Fig. 6.8). The arrangement of bands formed on

Table 6.2. Commonly used restriction enzymes with recognition sequence and source.

Enzyme	Recognition Sequence[†]	Microorganism
*Alu*I	AG▼C*T	*Arthrobacter luteus*
*Bam*HI	G▼GATCT	*Bacillus amyloliquefaciens H*
*Bgl*I	GCCNNNN▼NGCC	*Bacillus globigii*
*Bgl*II	A▼GATCT	*Bacillus globigii*
*Eco*RI	G▼AA*TTC	*Escherichia coli* RY13
*Eco*RII	▼CC*WGG	*Escherichia coli* R245
*Eco*RV	GA*T▼ATC	*Escherichia coli* J62P7G74
*Hae*II	RGCGC▼Y	*Haemophilus aegyptius*
*Hae*III	GG▼C*C	*Haemophilus aegyptius*
*Hind*III	A*▼AGCTT	*Haemophilus influenzae* R_d
*Hpa*II	C▼C*GG	*Haemophilus parainfluenzae*
*Msp*I	C*C▼GG	*Moraxella species*
*Pst*I	CTGCA*▼G	*Providencia stuartii* 164
*Pvu*II	CAG▼C*TG	*Proteus vulgaris*
*Sal*I	G▼TCGAC	*Streptomyces albus* G
*Taq*I	T▼CGA*	*Thermus aquaticus*
*Xho*I	C▼TCGAG	*Xanthomonas holcicola*

[†]Only one strand of the recognition sequence is given. ▼ denotes the cleavage point. The asterisk (*) indicates a modified base (A* is N^6-methyladenine and C* is 5-methylcytosine). N = any nucleotide, W = A or T, R = A or G, Y = T or C. Source: R. J. Roberts & D. Macelis, REBASE – restriction enzymes and methylases, Nucl. Acids Res. 21, 3125–3127 (1993).

the gel by electrophoresis is called a restriction map. Particular sequences can be located within a chromosome and the degree of differentiation between related chromosomes can be approximated using this map.

After fragmenting the DNA molecules, denaturing the fragments into ssDNA and separating these ssDNA via electrophoresis, sequencing of the bases within the fragments can be commenced.

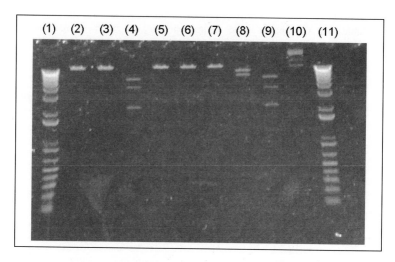

Fig. 6.8. Gel electrophoresis of a plasmid DNA with 14 kb. The two outer lanes, (1) and (11), are ladders of known length. The restriction enzymes used from left to right are: (2) BamH I, (3) EcoR I, (4) HinD III, (5) Not I, (6) Xho I, (7) EcoR I and BamH I, (8) Not I and EcoR I, (9) Not I AND HinD III, and (10) uncut DNA. The restriction enzyme Not I cuts the circular plasmid DNA molecule once and linearises it. (Courtesy of A. Ahern Department of Biological Sciences, Imperial College London, UK.)

6.3.2 *The Chemical Cleavage method (The Maxam–Gilbert method)*

The *chemical cleavage method* for DNA sequencing is also referred to as *Maxam–Gilbert method*. Sequencing of the ssDNA fragment starts by *radioactively labelling* the 5′-end with a ^{32}P atom. This is achieved by reacting the DNA molecule with $[\gamma\text{-}^{32}P]ATP$. The reaction is catalysed by an enzyme called *polynucleotide kinase*. If the DNA fragment already has a 5′-phosphate group, this must be removed first, by treatment with *alkaline phosphatase* (Fig. 6.9).

The radioactively labelled DNA fragment is then treated with a reagent that specifically cleaves the DNA molecule at a particular type of nucleotide. For example, a solution of hydrazine cleaves a DNA molecule before every *C*-nucleotide. The cleaving reaction is carried out such that low yields are obtained. The aim is not to cut all DNA molecules at locations containing a *C*-nucleotide. The target is to cut each molecule only once at a randomly located *C*-nucleotide within the chain.

Assuming the DNA fragment to be sequenced is

$$^{32}P\text{-ACCTGACATCG,}$$

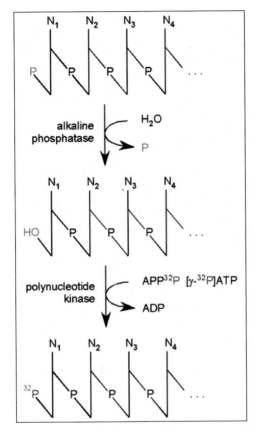

Fig. 6.9. ^{32}P labelling in the Maxam–Gilbert method: The 5′-end reacts with [γ-^{32}P]-ATP under catalysis of the enzyme polynucleotide kinase. If a phosphate group is present at the 5′-end, this must be removed first by treatment with alkaline phosphatase.

then cleavage of the 5′-side of the *C*-residues results in the following 5′-labelled fragments:

^{32}P-ACCTGACAT (9 bases)
^{32}P-ACCTGA (6 bases)
^{32}P-AC (2 bases)
^{32}P-A (1 base)

In case of a high yield reaction, cleavage would occur before every *C* base and the only resulting 5′-labelled fragment would be ^{32}P-A, rendering the technique useless.

The obtained fragments are separated according to their size by SDS-PAGE (section 3.2). The relative position on the gel can be determined by detecting the ^{32}P signal with *autoradiography*. To achieve unambiguous results, the resolving power of the gel must be good enough to resolve fragments that differ by only one base.

For full sequencing of a DNA molecule, the sample is divided into four aliquots. Each of these aliquots is subjected to a different chemical treatment in order to achieve base specific cleavage. The reaction products are then separated in parallel on a single gel and the sequence of bases can be identified.

Aliquot 1   Cleavage at G only

The DNA sample is treated with dimethyl sulfate (DMS), which results in methylation of the *G* residues at the *N*7 position. The glycoside bond of the methylated *G* residue can then be hydrolysed and the *G* residue is eliminated. In the next step, piperidine is added, which reacts with the hydrolysed sugar residue. This leads to the cleavage of the backbone (Fig. 6.10).

Aliquot 2   Cleavage at G and A

DMS also methylates *A* residues at their *N*3 position. Hence, treatment with piperidine also leads to cleaving of *A* residues. The rate of this reaction is only one fifth of that for the cleavage of *G* residues. However, if an acid is added to the reaction mixture instead of DMS, then both *A* and *G* residues are cleaved at a comparable rate. The positions of the *A* residues is determined by comparing the positions of the *G* and *G* + *A* residues.

Aliquot 3   Cleavage at C and T

Treatment of DNA with hydrazine and subsequent reaction with piperidine releases both *C* and *T* residues as shown in Fig. 6.11.

Aliquot 4   Cleavage at C only

If the reaction with hydrazine is carried out in a 1.5 M NaCl solution, then DNA is only cleaved before the C residues. Again comparison of the *C* + *T* and *C* positions reveals the positions of the *T* residues.

In all four reactions, the conditions are adjusted such that residues are released at an average of only one randomly located position per DNA molecule.

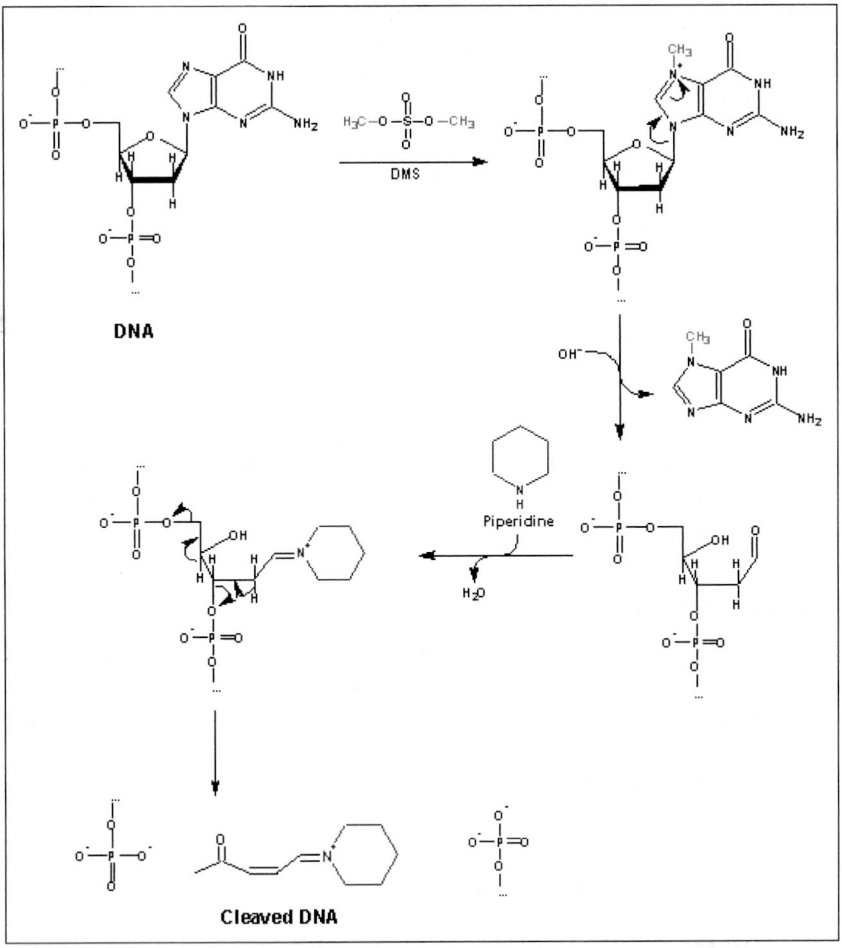

Fig. 6.10. The cleavage reaction for *G* residues in the Maxam–Gilbert method. If residues are protonated rather than methylated, cleavage will occur before both *G* and *A* residues.

As an example, the expected sequencing result of the DNA molecule ^{32}P-ACCTGACATCG is shown in Fig. 6.12.

It has to be noted, that the residue directly attached to the 5'-terminus cannot be identified, as the corresponding nucleotide is destroyed in the reaction. Moreover, the second residue is often impossible to resolve by the gel. The first and second residue must be identified by sequencing the complementary strand. This also serves as a verification of the initial sequence.

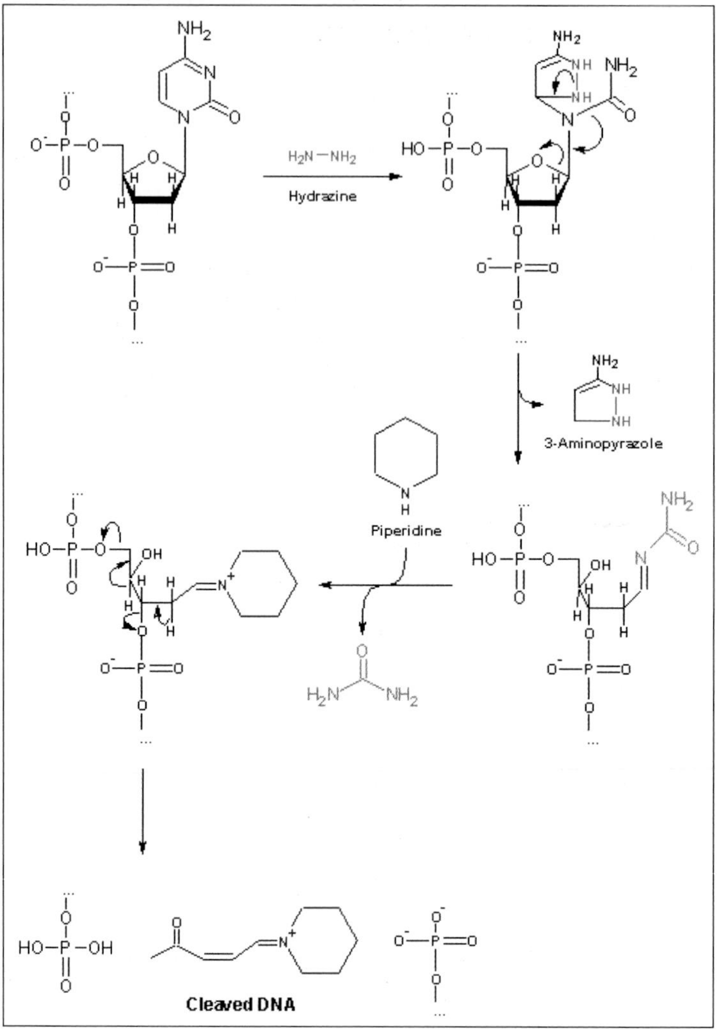

Fig. 6.11. The reaction used to cleave before *C* and *T* residues in the Maxam–Gilbert method. If the reaction is carried out in a solution containing 1.5 M NaCl, cleavage occurs before *C* residues only.

6.3.3 The Chain Terminator Method (The Sanger or Dideoxy method)

The *chain terminator method* is also referred to as *Dideoxy method* or *Sanger method*. It is based on synthesising a complementary strand of DNA along the

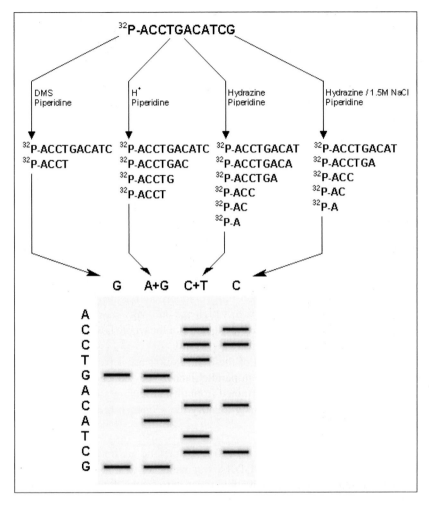

Fig. 6.12. Schematic for sequencing of a DNA fragment, using the Maxam–Gilbert method.

target DNA molecule, in a similar fashion to the PCR described in section 6.2. Deviating from the PCR procedure, the synthesis of this complementary strand is carried out in the presence of a chain terminating nucleotide, which is the key element for the determination of the base sequence.

The procedure of the dideoxy method starts by dividing the ssDNA sample into four aliquots. Each aliquot is incubated with DNA polymerase, the four dNTPs and a suitable primer. The complementary strands are labelled by incorporating an $[\alpha\text{-}^{32}\text{P}]$ into either the primer or into at least one dNTP. Additionally, to each of the

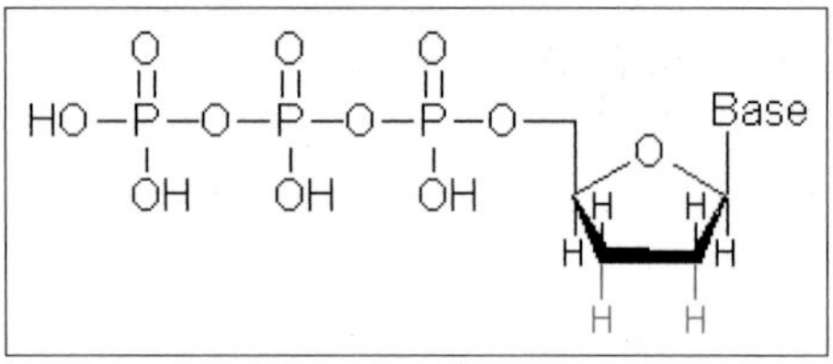

Fig. 6.13. The structure of $2',3'$-dideoxynucleotide triphosphate.

four separate reaction vessels, a small amount of one of the $2',3'$-*dideoxynucleotide triphosphates* (*ddNTP*) is added (Fig. 6.13).

As soon as a ddNTP is incorporated into the growing polynucleotide, the reaction is terminated, as there is no free $3'$-hydroxyl group for further incorporation of a nucleotide. Thus, the addition of ddNTPs leads to a series of truncated chains, each terminated by the dideoxy analogue in the position of the corresponding base.

It has to be noted that only very small amounts of each ddNTP are used in order to ensure that the probability of incorporating a ddNTP is very small. If ddNTP was used in large amounts, then the probability of incorporating the chain terminator at the first possible position would be very high and only very short fragments would be obtained. No information about the sequence of the whole DNA molecule could be obtained from these short copies.

The reaction products of each of the four aliquots are separated according to their size by gel electrophoresis in parallel lanes. The sequence of the complementary DNA strand can be determined from the autoradiogram (Fig. 6.14). This can then be used to unambiguously identify the sequence of the original DNA molecule.

Both the chemical cleavage and the dideoxy method are invaluable techniques that allow for the sequencing of DNA fragments up to 800 nucleotides long. Although the chemical cleavage method is easier to set up, the dideoxy method is generally preferred for routine use, as it is more readily automated. Commercially available sequencing instruments use one of the following strategies:

(1) Four separate reaction mixtures are set up according to the Sanger method procedure as described above. The primers are covalently bonded to a fluorescing dye at the $5'$-terminus. The reaction products are separated by gel electrophoresis in parallel lanes and the gels are read using a *laser induced fluorescence (LIF)* detector.
(2) The primers in each reaction vessel are labelled with a different fluorescent dye. After the reaction, the mixtures are combined and are separated in a single lane by gel electrophoresis. The terminal base of each fragment is identified by the characteristic fluorescence spectrum (Fig. 6.15).

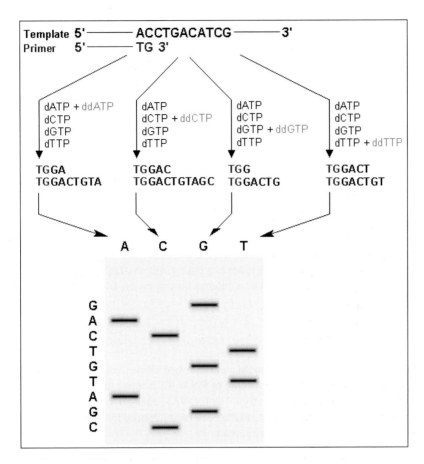

Fig. 6.14. Schematic of the expected sequencing result for the DNA fragment (^{32}P-ACCTGACATCG) using the Sanger method.

(3) In a single vessel, the reaction is set up using ddNTPs each covalently bonded to a different fluorescent dye. The resulting products are sequenced in a single lane and the terminal base is identified by the characteristic fluorescence of the attached dye.

High throughput instruments have emerged by combining the chain terminator method with computerised procedures and robotics. Moreover, developments in electrophoretic separations, and more specifically CE, have minimised the separation time. Such advances have enabled the sequencing of the complete genome of species including the recently completed human genome project.

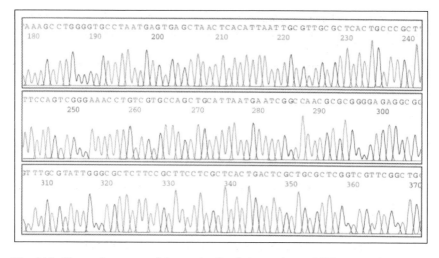

Fig. 6.15. Electropherogram of the result of a chain terminator DNA sequencing experiment using primers each labelled with a different fluorescent dye (A = green; C = blue; G = black, T = red).

6.4   RNA Sequencing

The techniques for DNA sequencing as described above can also be used, with slight modifications for the sequencing of RNA. RNA molecules must initially be transcribed to their complementary DNA (cDNA) sequence by reverse transcription (section 6.3.5). The resulting cDNA can be sequenced by either the chemical cleavage or the chain terminator method.

Summary

Prior to any analysis, the nucleic acids must be isolated from the cells. The cell walls are lysed and the DNA or RNA molecules can be extracted from the released cell contents and cell debris. Methods employed for this include precipitation, liquid–liquid extraction, chromatography and centrifugation. Enzymes are often used to degrade proteins and other undesired molecules. The isolated DNA molecules can then be amplified using the polymerase chain reaction (PCR). In order to amplify RNA using PCR, it is necessary to synthesise cDNA along the RNA strand by reverse transcription. Sequencing can only be performed with ssDNA fragments shorter than 800 bases. Controlled fragmentation of large DNA molecules can be achieved by treatment with restriction enzymes. The obtained dsDNA fragments are then denatured to give ssDNA fragments, which are separated from each other by gel electrophoresis. DNA sequencing can be performed by the chemical cleavage method (Maxam-Gilbert method) or by the chain terminating method

(Sanger method). For RNA sequencing, a complementary cDNA has to be synthesised by reverse transcription. This cDNA can then be sequenced by the same methods as ordinary DNA.

References

1. C. R. Newton and A. Graham, *PCR*, Bios Scientific Publishers, 1997.
2. M. J. McPherson, B. D. Hames and G. R. Taylor (editors), *PCR: A Practical Approach*, IRL Press, 1995.
3. K. B. Mullis, F. Ferré and R. A. Gibbs (editors), *PCR: The Polymerase Chain Reaction*, Birkhäuser, 1994.
4. L. Alphey, *DNA Sequencing, from Experimental Methods to Bioinformatics*, Bios Scientific Publishers, 1997.
5. D. Voet and J. G. Voet, *Biochemistry*, 2nd edition, Wiley and Sons, 1995.

Chapter 7

PROTEIN SEQUENCING

In this chapter, you will learn about...

♦ ... the strategies for protein sequencing.
♦ ... how a protein can be unfolded prior to sequence analysis.
♦ ... how the amino acid composition of a protein can be determined.
♦ ... how the amino groups at the *C*-terminus and *N*-terminus can be identified.
♦ ... reactions for cleaving specific peptide bonds in a protein to generate fragments.
♦ ... sequencing of protein fragments using Edman degradation.
♦ ... how disulfide bridges can be located within a protein.
♦ ... and how the complete protein structure can be determined from individually sequenced polypeptide fragments.

The objective of protein sequencing is to determine its primary structure, i.e. the sequence of the amino acids in the polypeptide chain. This primary sequence must be identified before elucidation of the secondary, tertiary and quaternary structures (section 1.1.2.2) can begin. The primary structure of a protein is essential for the protein's molecular function and mechanism. Genetic mutations often lead to changes in the amino acid sequence of a protein. In certain cases this can lead to malfunction and disease. Identifying and recognising such changes may help in developing diagnostic tests or even symptom relieving therapies.

In this chapter, the *strategies* for identifying the amino acid sequence in a polypeptide chain are outlined. The *amino acid composition* of a protein, i.e. the relative abundance of each amino acid in the peptide chain is often a first step to protein identification. This however, does not give any information about the sequence of the amino acids. Methods for determining which amino acid is at the *N- and C-terminus* of the polypeptide chain are described, as well as *degradation reactions* that specifically cleave disulfide bridges or selectively break certain peptide bonds within the protein. The amino acids of the fragments obtained can then be sequenced using *Edman degradation*. Similar to a jigsaw, the complete protein structure can finally be put together.

7.1 Protein Sequencing Strategy

Amino acid sequencing was pioneered by Frederick Sanger. In 1953, he found the sequence of bovine insulin, a polypeptide consisting of 51 amino acid residues (Fig. 1.19). In 1958, Sanger was awarded the Nobel Prize for his contributions to determine the structure of insulin and other proteins. Despite the technological advances of the past decades, the basic strategy for revealing the primary structure of a protein, which was originally developed by Sanger, is still being used today.

This strategy can be divided roughly into the following steps:

- The number of distinct polypeptide chains (*subunits*) in the protein must first be determined.
- *Disulfide bonds*, which can occur along a single polypeptide chain or between different polypeptide chains, must be cleaved.
- The *amino acid composition* of each polypeptide chain can then be established.
- The subunits are often too long to be sequenced directly. Hence, they must be fragmented into sets of smaller peptides by *specific cleavage reactions*.
- The sequence of each of these fragments is then uncovered by employing *Edman degradation*.
- The sequence of the complete subunit can then be put together by comparing overlaps of the different sets of fragments.
- Finally, the structure of the whole protein including the disulfide bridges between different subunits can be determined.

These steps are discussed in more detail in the following sections.

7.2 End-group Analysis

Since each polypeptide contains an *N*-terminal and a *C*-terminal residue, the number of distinct subunits in a protein can be determined by identifying the number of either of these end groups, provided the subunits are not circular and the end groups are not chemically protected.

7.2.1 *N-terminal Analysis (Edman Degradation)*

The *N*-terminus of a polypeptide can be determined by reaction with *dansyl chloride*, with *Edman's reagent* or with an *aminopeptidase*.

1-dimethyl aminophthalene-5-sulfonyl chloride (dansyl chloride) reacts with primary amine groups in the polypeptide chain and forms a dansyl polypeptide. Upon acidic hydrolysis, all peptide bonds in the chain are cleaved and the *N*-terminal residue of the peptide is liberated in the form of a dansyl-amino acid (Fig. 7.1). This amino acid derivative is highly fluorescent and can thus be detected with very high sensitivity.

A particularly useful method for *N*-terminal analysis is the *Edman degradation*. The *N*-terminal amino acid of a polypeptide is reacted with *Edman's reagent* (phenyl isothiocyanate, PITC) under mild alkaline conditions to produce a phenylthiocarbonyl polypeptide adduct (Fig. 7.2). This adduct is then treated with a strong acid such as anhydrous trifluoroacetic acid. The *N*-terminus of the peptide chain is cleaved as a thiazolinone derivative, without breaking any other peptide bonds in the chain. The thiazolinone amino acid is then selectively extracted into an organic solvent and treated with an acid to form the more stable phenylthiohydantoin (PTH) derivative. The obtained PTH amino acid can be detected using UV-absorption ($\lambda_{max} = 296$ nm). It is separated from the other components by chromatography or electrophoresis. The amino acid contained in the PTH derivative can be identified either according to its retention time or according to its mass, by coupling the separation method to a mass spectrometer (section 4.3.4).

In contrast to other processes for *N*-terminal analysis, Edman degradation only leads to cleavage of the *N*-terminal residue, whilst leaving the remaining polypeptide intact. The reaction can be repeated in a cyclical fashion and each time the *N*-terminus is cleaved, extracted into the organic phase and identified. With this process, relatively short peptides with 40 to 60 residues can be sequenced (section 7.7).

Sequential analysis of *N*-termini can also be achieved by *enzymatic reactions*. An exopeptidase cleaves a terminal polypeptide residue. An aminopeptidase, for example, cleaves *N*-terminal residues (Fig. 7.3a). Aminopeptidases, however, have only a limited use for the determination of an amino acid sequence. Due to their high specificity, only selected amino acids are cleaved and these at different rates. Some residues are not cleaved by certain aminopeptidases, in which case the sequencing experiment stops. Some amino acid residues may be more resistant to the enzyme than others. The different cleavage rates make unambiguous sequence determination difficult.

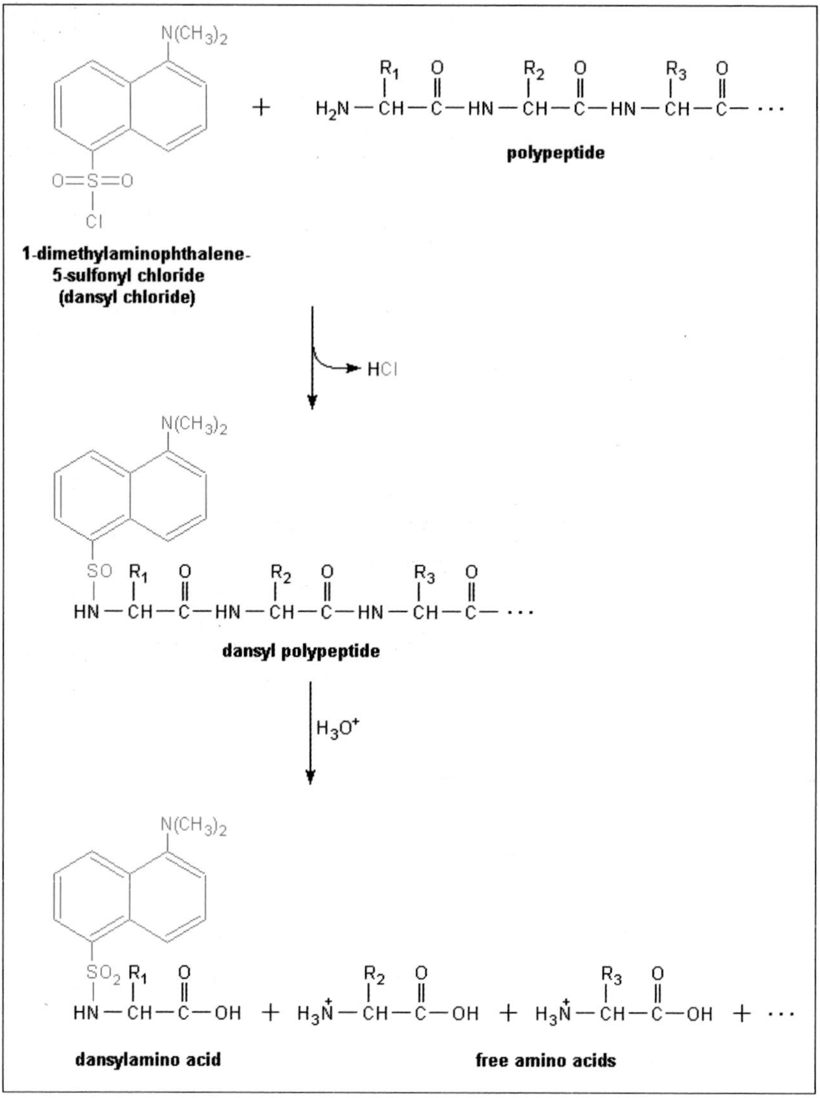

Fig. 7.1. The reaction of dansyl chloride with a polypeptide and subsequent hydrolysis yields a fluorescent derivative of the *N*-terminal residue.

7.2.2   *C-terminal Analysis*

Another class of exopeptidases, the *carboxypeptidases*, can be used for the *C*-terminal analysis of polypeptides. Carboxypeptidases, in direct analogy to

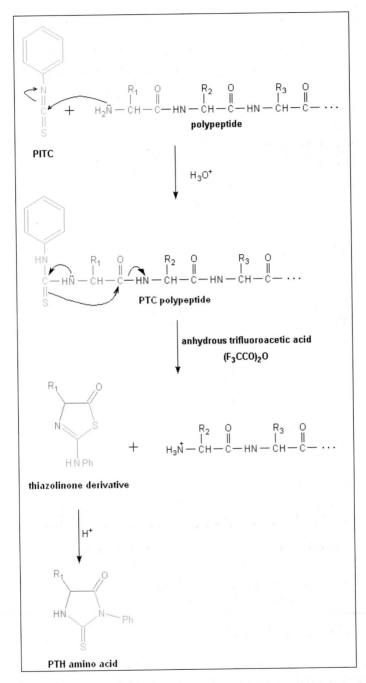

Fig. 7.2. The Edman degradation is a powerful process for protein sequencing as it allows for sequentially removing the *N*-terminal amino acid residue in repeated and controlled steps.

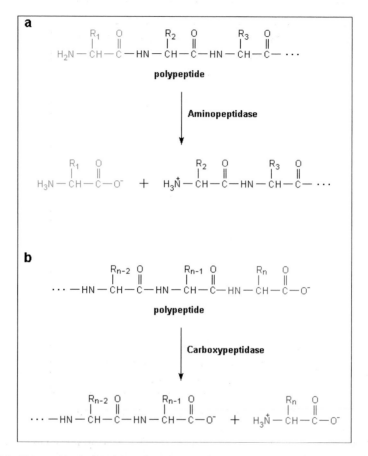

Fig. 7.3. Exopeptidases cleave terminal residues from a peptide chain. (a) Treatment with aminopeptidases leads to cleavage of the *N*-terminal residue, whereas treatment with (b) carboxypeptidases results in breaking off the *C*-terminal residue.

aminopeptidases, cleave the *C*-terminal residue of the polypeptide (Fig. 7.3b). Again, the selective nature of the enzymes means that certain residues are resistant or cleaved at slow rates.

Alternatively, *chemical methods* can be employed, for example *hydrazinolysis*. The polypeptide is treated with anhydrous hydrazine at 90 °C for 20-100 h under mild acidic conditions. This reaction produces aminoacyl hydrazine derivatives of all amino acid residues, except for the *C*-terminal residue, which is released as a free amino acid (Fig. 7.4). After chromatographic separation of the reaction mixture, this free amino acid can be identified. Hydrazinolysis also leads to a great number of side products.

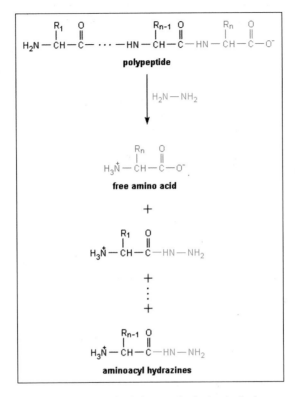

Fig. 7.4. *C*-terminal cleavage by hydrazinolysis.

There is no reliable process comparable to the Edman degradation for sequential *C*-terminal analysis. None of the described methods can be employed in a cyclic fashion to extract one amino acid after the other and identify them in an automated fashion.

7.3 Disulfide Bond Cleavage

Disulfide bonds (section 1.1.2.2) are formed between two Cys residues. These sulfide bridges can occur within a polypeptide chain or between different poly-peptide chains, i.e. between different subunits. The S–S bonds must be cleaved prior to sequencing to separate and unfold the subunits. Disrupting the protein's native conformation also facilitates the action of the proteolytic agents used for sequencing (section 7.7). The cleaving reactions are best carried out under *dena-turing conditions*, for example by adding guanidine hydrochloride or detergents such as SDS. The compact protein structure becomes unfolded and all disulfide bonds are exposed.

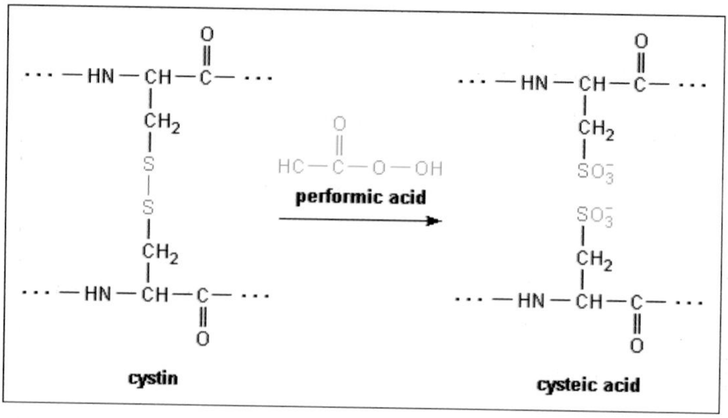

Fig. 7.5. Oxidative cleavage of the disulfide bonds by performic acid.

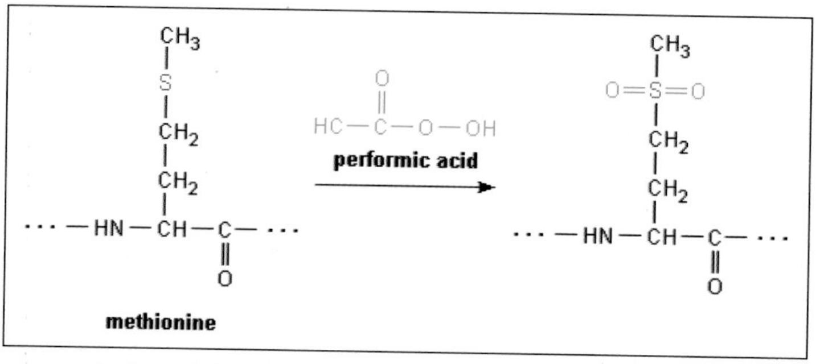

Fig. 7.6. Oxidation of methionine residues by performic acid.

Disulfide bonds may be broken by *oxidation with performic acid*. All Cys residues in the protein, whether they are linked by sulfide bridges or not, are oxidised to cysteic acids (Fig. 7.5). The total Cys content can be deduced from the amount of cysteic acid produced.

There are, however, significant disadvantages associated with this method as the performic acid also reacts with other groups in the polypeptide chain. For example, it partially destroys the indol side chain of Trp and it oxidises Met residues (Fig. 7.6). This inhibits the specific cleavage of Met in later sequencing steps (section 7.7).

Alternatively, sulfide bonds can be *reduced* to thiols by dithiothreitol (DTT) or 2-mercaptoethanol (Fig. 7.7). DTT, also referred to as Cleland's reagent, is more widely used. The resulting thiol (–SH) groups must be treated with an *alkylating agent* such as iodoacetic acid to prevent the reformation of disulfide bonds (Fig. 7.8). These *S*-alkyl derivatives are also stable under the peptide cleaving conditions employed in subsequent steps.

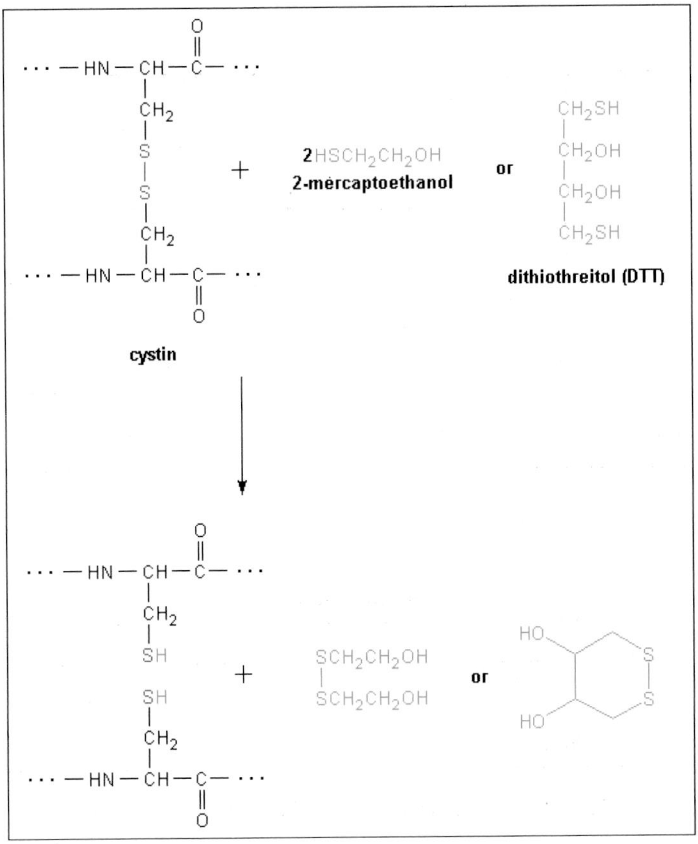

Fig. 7.7. Reductive cleavage of disulfide bonds by 2-mercaptoethanol or Cleland's reagent.

7.4 Separation and Molecular Weight Determination of the Subunits

After cleavage of S–S bridges, the individual subunits of the protein are *separated* by electrophoretic methods such as SDS–PAGE (section 3.2.3) or by chromatographic methods such as SEC (section 2.3.4) or RP-HPLC (section 2.3.1).

From the *molecular weight* of each subunit, the number of amino acid residues can be determined, as each amino acid residue has a mass of about 110 Da. Traditionally, molecular weights of proteins were determined by SDS–PAGE (section 3.2.3) or SEC (section 2.3.4). However, mass spectrometry (chapter 4) is now routinely used, as it is often a far more accurate and faster method.

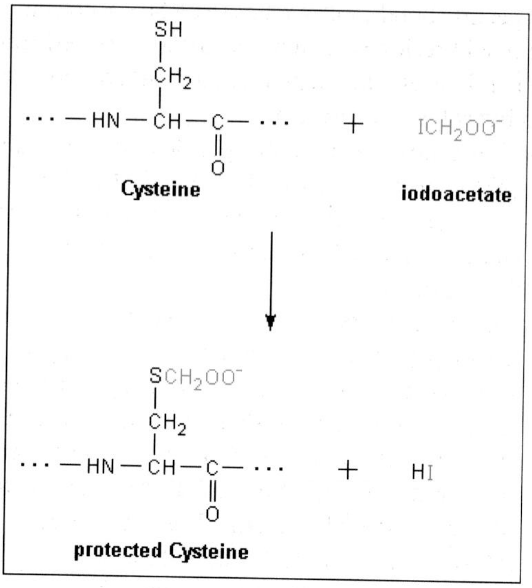

Bioanalytical Chemistry

Fig. 7.8. The –*SH* groups are protected by alkylation with iodoacetic acid to prevent re-oxidation.

7.5 Amino Acid Composition

The amino acid composition, i.e. the quantity of each amino acid residue in the peptide chain, is a characteristic parameter for each protein. Often, an unknown protein can be identified by measuring the relative percentage of the amino acid residues and comparing these to a database.

Measurement of the amino acid composition can be achieved in two steps. First, all the peptide bonds in the protein are cleaved by either acidic, basic or enzymatic hydrolysis. Subsequently, the free amino acids are separated from each other and quantified.

In *acid catalysed hydrolysis*, polypeptides are treated with 6M HCl under vacuum to prevent oxidation of the sulfur containing amino acids by air. The reaction mixture is heated to $100-120\,^\circ$C for about 24 h. Longer reaction times may be required for complete hydrolysis of the aliphatic amino acids Leu, Val and Ile. However, some other amino acid residues are degraded under these harsh conditions. The rate of degradation of certain residues such as Thr, Tyr and Ser can be measured and a correction factor can be included to account for their loss as a function of the hydrolysis time. Trp, however, is extensively degraded. Furthermore, acidic hydrolysis converts Asn to Asp and Gln to Glu, respectively; NH_4^+ is eliminated in both reactions. To determine the amount of these amino acids, the total content of Asx (Asn + Asp), Glx (Gln + Glu) and NH_4^+ (Asn + Gln) must be measured and compared. Optimum hydrolysis conditions are difficult to establish, as total peptide bond cleavage has to be balanced against degradation of some of the amino acid residues. Often, the reaction is performed several times under different conditions and the actual amino acid composition is extrapolated from the different hydrolysis experiments.

Base-catalysed hydrolysis is even more problematic and only used in special cases. The polypeptides are reacted with 4 M NaOH at 100 °C for 4 to 8 h. Under these conditions Arg, Cys, Ser, and Thr are decomposed and other amino acids are de-aminated and racemised. This limits the use of basic hydrolysis to the determination of the Trp content.

Enzymatic methods are most often used for determining the Asn, Gln and Trp content. Exo- and endopeptidases are enzymes that catalyse the hydrolysis of specific peptide bonds (section 7.6.1). As these enzymes exhibit a high specificity, it is essential to use a mixture of them to ensure hydrolysis of all peptide bonds. Enzymes should be used at low concentrations ($\sim$1 %) as they are proteins themselves that can degrade and thus contaminate the reaction mixture.

None of the hydrolysis methods described above can be used on its own to achieve total hydrolysis of all peptide bonds without degrading any amino acid residues. Any of the methods can be used to quantify certain amino acids. By combining two or three hydrolysis methods, all amino acids in the polypeptide can be quantified.

After hydrolysis, the free amino acids are separated by ion exchange chromatography (section 2.3.2) or RP-HPLC (section 2.3.1). They can be identified according to their elution times and quantified according to their elution volumes. To increase sensitivity, the amino acids are usually derivatised either pre- or post-column. Dansyl Chloride, Edman's reagent (Fig. 7.1) as well as orthophthalaldehyde (OPA) and 2-mercaptoethanol can be employed to form highly fluorescent adducts (Fig. 7.9) that can be detected easily.

7.6   Cleavage of Specific Peptide Bonds

Direct sequencing can only be achieved for peptides no longer than about 50 residues. Beyond this length, results become unreliable due to incomplete reactions and the accumulation of impurities from side reactions. Hence, most proteins must be cleaved into smaller fragments prior to sequencing. A number of chemical and enzymatic reactions are available that break peptide bonds within the chain at specific places.

The aim of these cleavage reactions is to generate a number of fragments that are small enough to be sequenced. Often the protein sample is divided into two aliquots. These are fragmented with different agents leading to two different sets of fragments. After sequencing, the fragments can be ordered due to partial overlap (section 7.8).

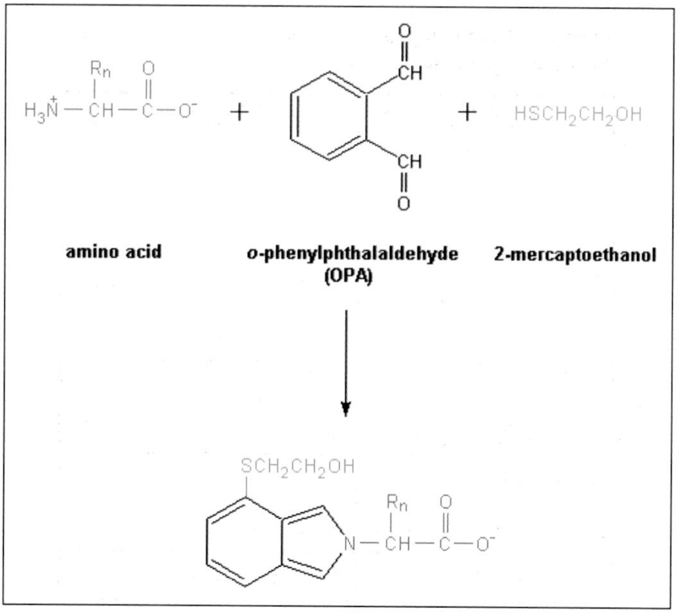

Fig. 7.9. Derivatisation of amino acids by OPA and 2-mercaptoethanol leads to highly fluorescent adducts.

7.6.1   Enzymatic Fragmentation

Some of the proteolytic enzymes that are used for specifically fragmenting proteins are listed in Table 7.1. They can be separated into exo- and endopeptidases. An *exopeptidase* cleaves either the *N*- or *C*-terminal amino acid from the peptide chain (section 7.2). An *endopeptidase* cuts a peptide at specific locations within the chain to generate a number of fragments that are characteristic for each protein.

Trypsin is the most commonly used endopeptidase, due to its high specificity. It catalyses the cleavage of the peptide chain at the carboxyl side (*C*-side) of positively charged residues (Arg and Lys), but only if the next residue is not Pro (Fig. 7.10). Tryptic digestion is often used as a stand-alone identification method for known proteins. The characteristic fragment pattern obtained can be compared to a database. This is often sufficient to identify a protein.

To take further advantage of trypsin's specificity, *cleavage sites* may be *removed or added*. If the positive charge in the Lys or Arg side chain is eliminated, trypsin no longer cuts the peptide at this point. Additional cleaving sites may be generated by introducing a positive charge into side chains of other amino acids. For example, derivatising a Lys residue with citraconic anhydride leads to a Lys side chain that is no longer positively charged and, therefore, not recognised by trypsin as

Table 7.1. Commonly used proteolytic enzymes for partial digestion of proteins into smaller fragments.

Enzyme	Specificity
Endopeptidases	
Trypsin	$R_{n-1} = $ Arg, Lys $R_n \neq$ Pro
Pepsin	$R_n = $ Leu, Phe, Trp, Tyr, Val $R_{n-1} \neq$ Pro
Chymotrypsin	$R_{n-1} = $ Phe, Trp, Tyr $R_n \neq$ Pro
Endopeptidase GluC	$R_{n-1} = $ Glu
Exopeptidases	
Leucine aminopeptidase	$R_1 \neq$ Pro
Aminopeptidase *M*	all *N*-terminal residues
Carboxypeptidase *A*	$R_n \neq$ Arg, Lys, Pro $R_{n-1} \neq$ Pro
Carboxypeptidase *B*	$R_n = $ Arg, Lys $R_{n-1} \neq$ Pro
Carboxypeptidase *C*	all *C*-terminal residues

a cleavage site (Fig. 7.11). After tryptic digestion, this functionality can be removed under mild acidic conditions. On the other hand, Cys can be derivatised with a β-haloamine such as bromoethylamine. This aminoalkylation reaction introduces a positive charge into the Cys side chain and thus produces a new cleavage site for trypsin (Fig. 7.12).

Endopeptidases that are less specific than trypsin may also be used. However, the obtained fragments may be too small and might not exhibit enough overlap with other fragments to enable ordering of the fragments into the right sequence (section 7.8).

By adjusting the reaction conditions and shortening the reaction time, the enzymes have only a limited access to the polypeptide due to steric hindrance. This approach is known as *limited proteolysis*. The method is also useful for trypsin digestion of proteins with a high Arg and Lys content.

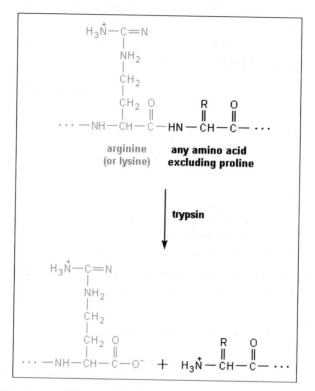

Fig. 7.10. Tryptic digestion of polypeptides.

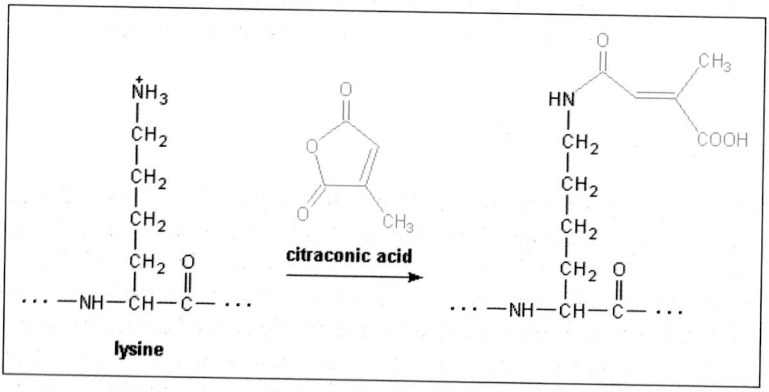

Fig. 7.11. Protection of Lys (and Arg) from tryptic digestion by reaction with citraconic acid.

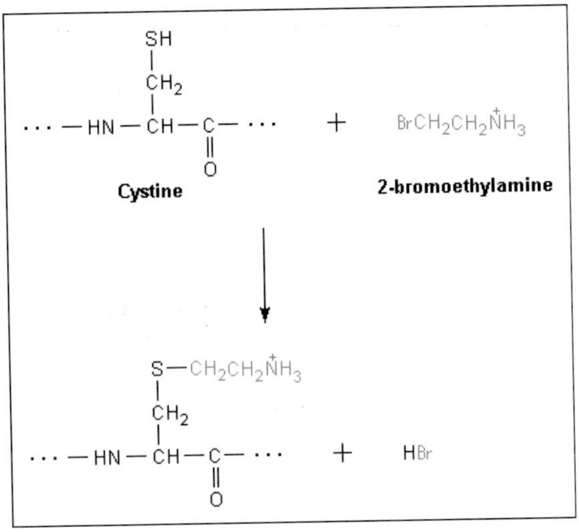

Fig. 7.12. Derivatisation of cystein with bromoethylamine. The introduction of a positive charge makes cystein susceptible to tryptic digestion.

7.6.2 *Chemical Fragmentation Methods*

There are also a number of chemical methods to specifically cleave peptide bonds. For example, *cyanogen bromide (CNBr)* specifically cleaves Met residues at the *C*-side, forming a peptidyl homoserine lactone (Fig. 7.13). To ensure that all Met residues are cleaved, the reaction takes place under denaturing mildly acidic conditions.

If the resulting peptide fragments are still too long for sequencing, a further fragmentation step may be required using different cleavage processes.

7.7 Sequence Determination

After cleaving the polypeptide chains into sufficiently small fragments, these must be separated by chromatography or electrophoresis and sequenced individually. The method of choice is Edman degradation, as described in section 7.2.1.

Edman degradation can be carried out in a fully automated instrument known as *sequenator*. The core of the sequenator is the reaction chamber. The protein must be immobilised inside the chamber. This is commonly achieved by bonding the proteins into a solid support or by adsorbing it onto an inert glass frit. Controlled amounts of reagents are injected by a pumping system without dead-volumes. The thiazolinone derivative that is produced during the Edman degradation reaction is

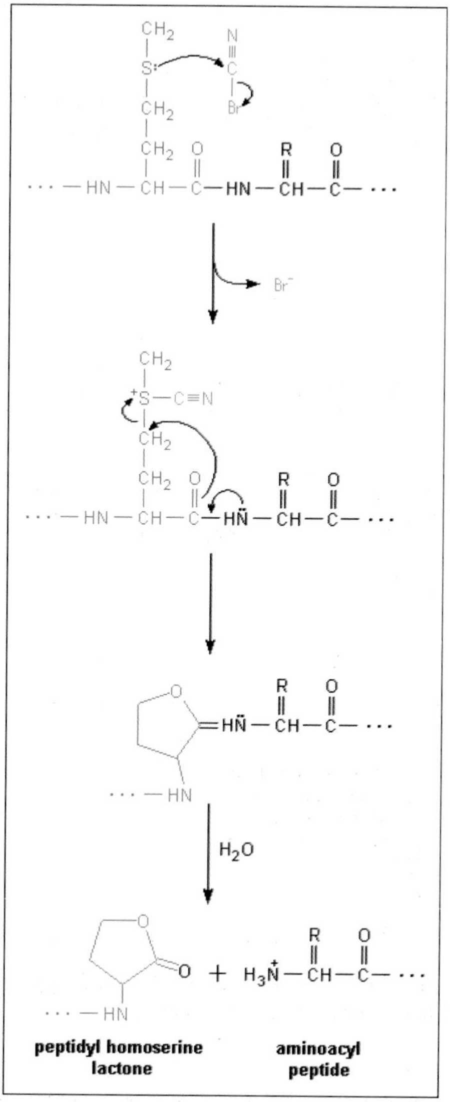

Fig. 7.13. Specific cleavage of methionine residues by CNBr.

then transferred into a conversion chamber, where hydrolysis to the PTH amino acid is carried out. This final product is pumped into an HPLC column for on-line analysis. Sequencing of a protein with 50 amino acid residues can be completed in less than an hour. The amount of protein required for unambiguous analysis is as low as pmol.

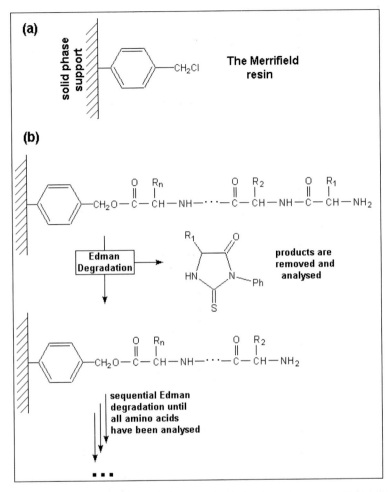

Fig. 7.14. (a) The Merrifield resin is commonly used as solid support matrix. (b) The protein is covalently bonded to the resin for Edman sequencing.

An example of a reaction on a *solid-phase* matrix, the Merrifield resin, is shown in Fig. 7.14. The peptide fragments are covalently bonded, either onto a polymer membrane or onto micrometer sized beads. The solid support is then immersed into a liquid phase and Edman degradation is carried out by sequentially adding the required reagents and removing the products for analysis.

For *gas-phase sequencing*, the sample is immobilised onto a chemically inert glass frit, often by using a carrier material such as polybrene. Reagents are carried in a gaseous stream of argon and delivered to the glass frit in minute

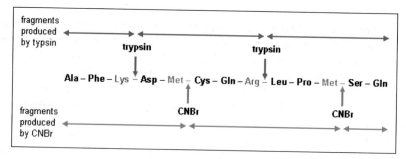

Fig. 7.15. Schematic of a polypeptide cleaved by trypsin and CNBr. Sites susceptible to tryptic cleavage are marked blue, while sites susceptible to CNBr cleavage are indicated with red colour. By comparing the amino acid sequences of the two sets of mutually overlapping fragments, the order of the fragments in the subunits can be deduced.

but accurately controlled quantities. Loss of protein due to mobilisation into acidic or basic solutions is prevented with this method. The reaction products are again automatically removed for chromatographic analysis and detected on-line.

7.8   Ordering of Peptide Fragments

Sequencing of the individual peptide fragments is followed by establishing the order, in which they were connected originally. This is achieved by comparing the amino acid sequence of one set of peptide fragments with the sequences of a second set of fragments, whose specific cleavage points differ from those of the first (Fig. 7.15).

Overlapping of fragments from different cleavage reactions must be long enough to uniquely identify the cleavage site. Given, however, that there are 20 possibilities for each amino acid position, an overlap of a few residues is usually sufficient.

7.9   Determination of Disulfide Bond Positions

To conclude the elucidation of the primary structure of a protein, the positions of possible disulfide bonds need to be determined. In order to do this, the native protein is cleaved as described in section 7.6. This results in a mixture of fragments, some of which are linked by disulfide bonds. The mixture is then subjected to 2*D* gel electrophoresis (section 3.2.5), using the same separating conditions in both dimensions. After the separation in the first dimension, the matrix is exposed to performic acid, which cleaves all possible disulfide bonds (section 7.3). Then the separation in the second dimension is carried out. Fragments that did not contain any S–S bridges, are positioned along the diagonal of the matrix, as their rate of migration is the same in both dimensions. The polypeptide fragments originally linked by a disulfide bond are oxidised by the performic acid and cleaved. The electropherogram shows two spots for the two fragments, positioned off the diagonal axis. The disulfide-linked fragments can be isolated from the gel and identified by sequencing as described above. Their amino acid sequence can then be compared to that of the whole protein and the location of a disulfide bond can thus be established.

7.10 Protein Sequencing by Mass Spectrometry

Advances in mass spectrometry, in particular advances in the ionisation techniques for biomolecules, in conjunction with automation and bioinformatics, have revolutionised the process of protein sequencing and minimised the analysis times. The sequencing of proteins by mass spectrometry is discussed in more detail in section 4.3.4.

Summary

Protein sequencing is a rather labour intensive process, which requires a large number of steps to be carried out. The protein must first be isolated from a complex cell matrix. The number of polypeptide chains (subunits) within the protein is determined next. This is followed by denaturing the protein, separating the subunits and cleaving them into fragments of about 50 amino acid residues. Usually, the subunit is divided into two aliquots and each is treated with a different cleaving agent to yield two sets of fragments. These fragments can then be sequenced. To elucidate the structure of the original subunit, the sequences of the two sets of fragments must be compared. Due to partial overlapping, the fragments can be put into the right order. Finally, the positions of disulfide-bridges can be identified.

Once a protein has been fully sequenced, it is usually saved in a publicly accessible database for on-line comparison. Researchers only have to determine the amino acid composition of their sample protein and compare this to the database. Very often, unambiguous identification of the sample protein is possible. Alternatively, the sample protein can be partially digested, for example by trypsin. The fragments of this tryptic digest are characteristic for a particular protein, in the same way as a fingerprint is characteristic for an individual. By comparison with an electronic database, unambiguous identification is often possible.

Protein sequencing is in many ways more complex than DNA sequencing. Proteins contain 20 different amino acids, which exhibit different chemical functionalities and reactivity. Every denaturing or cleaving reaction is accompanied by a number of side reactions that complicate the process. This is one of the reasons why only relatively short stretches of amino acids can be sequenced. The determination of the complete protein structure is further complicated by post-translational modifications that introduce a number of other compounds into the molecule. In contrast, DNA molecules consist of only four different bases, which have relatively similar chemical structures and properties, as well as a sugar component and phosphate groups. DNA sequencing of several hundred bases is routine. Furthermore, DNA molecules can be amplified by PCR. No similar method is available for proteins. Hence, every analysis method for proteins must be optimised towards high sensitivity, as the availability of the protein sample is often limited.

References

1. D. Voet and J. G. Voet, *Biochemistry*, 2nd edition, Wiley and Sons, 1995.
2. C. K. Mathews, K. E. van Holde and K. G. Ahern, *Biochemistry*, 3rd edition, Addison Wesley Longman, 2000.
3. S. Roe (editor), *Protein Purification Techniques: A Practical Approach*, Oxford University Press, 2001.
4. D. M. Webster (editor), *Protein Structure Prediction, Methods and Protocols*, Humana Press, 2000.
5. M. J. Bishop and C. J. Rawlings, *DNA and Protein Sequence Analysis: A Practical Approach*, Oxford University Press, 1996.

INDEX

2,2′-azino-bis
(ethyl-benzothioazoline-6-sulfonate)
123
2′,3′-dideoxnucleotide triphosphate 164
2D-GE
see 2-dimensional gel electrophoresis
2-dimensional gel electrophoresis 24, 25,
67ff
DNA 25, 69
fingerprint 67
proteins 26, 68
RNA 69
2-mercaptoethanol 64, 176, 179
3′-end 18
5′-3′ exonuclease activity 151
5′-end 18

α-amino acids 2
ab initio sequencing 156
Ab-Ag-complex
see antibody-antigen-complex
ABTS
see 2,2′-azino-bis
(ethyl-benzothioazoline-6-sulfonate)
α–C-atom 2
acetylation 20
acetylcholine esterase 122
Adenine 15, 18
adenosine-5′-phosphosulfate 137
adenosine triphosphate 138
affinity 110, 111, 114
affinity chromatography 34, 40
antibody 40
group-specific ligands 42
mono-specific ligands 42
non-specific desorption 40
specific desorption 40
stationary phase 40
Affymetrix 133, 136
agarose 40, 48, 59
aggregation number 79
α-helix 11
AIDS 124
albumin 113
alcohol oxidase 127
alcohol test 121
alkaline phosphatase 122, 158

alkylating agent 176
amino acid composition 25, 169, 178
acid catalysed hydrolysis 178
base catalysed hydrolysis 179
enzymatic methods 179
amino acids 1, 2
abbreviations 2, 4, 5
acidic 2
aliphatic 2
analysis 35, 37, 45, 78, 178
aromatic 2
basic 2
classification 2, 4
configuration 2
dissociation constants 5, 6
hydroxyl containing 2
isoelectric point 6, 7
negatively charged 6
non-polar 4
peptide bond 9
pI values 6, 7
pK values 5, 6
polar 6
polarity 4
positively charged 6
secondary 2
structure 2
sulfur containing 2
aminopeptidase 171
amperometric glucose sensor
see blood glucose sensor
ampholyte
see carrier ampholyte
amphoteric 6
annealing 146
anodic mobilisation 77
anolyte 66
antibody 8, 42, 110, 111, 127
affinity chromatography 42
as bioreceptors 127
bivalent 112
F(ab)$_2$-fragment 113
Fab-fragment 112, 113
Fc-fragment 112, 113
fragments 112, 113
hosts 111
monoclonal 112

189

antibody (*cont.*)
  paratope 110, 112, 114
  polyclonal 112
  primary 117, 118
  secondary 117, 118
  structure 111, 112
antibody-antigen complex 110, 114
  affinity 114
  avidity 115
  equilibrium constant 115
antigen 8, 42, 110, 113
  complete 113
  epitope 110, 113, 114
  hapten 113
  incomplete 113
  multi-determinant 113
antigenic 113
apparent mobility 54
APS
  see adenosine-5′-phosphosulfate
apyrase 137, 138, 139
assay formats 115
  competitive 115
  excess reagent 116
  heterogeneous 119
  homogeneous 119
  labelled 115
  limited reagent 115
  non-competitive 116
  sandwich assays 117, 118
  unlabelled 115
ATP sulfyrase 137, 138
autoradiogram 160
autoradiography 160
avidity 115

backbone
  nucleic acids 18
  proteins 10
base pairs 18, 19
β-D-deoxyribose 15, 16
β-D-ribose 15, 16
β-folding 12
bioassay formats
  see assay formats
bioassays 25, 110
  detection 119
  dose-response curve 116, 118
  pregnancy test 120
biomolecular recognition
  see molecular recognition
bioreceptor 125, 126
biosensor 24, 125
  applications 125, 126
  bioreceptor 125, 126

blood glucose 128
  operating principle 129
  transducer 125, 127
bivalent 112
blood glucose 121, 126, 128
blood glucose sensor 128
blood type 121
blotting 63, 69
β-mercaptoethanol *see* 2-mercaptoethanol
bovine milk
  MALDI-TOF-spectrum 96
bovine serum albumin 97, 152
  MALDI-TOF-spectrum 97
bromoethylamine 181
BSA
  see bovine serum albumin
β-sheet
  see β-folding
buffer
  control of EOF 53
  for chromatography 35, 39, 40, 45
  for electrophoresis 50, 53, 61, 72, 79
  for ESI-MS 102
  for MEKC 79
  for PCR 152
buoyant density 144

capacity factor 32
  in MEKC 81
capillary electrophoresis 24, 25, 69
  amino acids 69, 78
  background electrolyte 72
  band broadening 72
  buffers 72
  capillaries 72
  capillary gel electrophoresis 82
  capillary isoelectric focussing 70, 76
  carrier electrolyte 72
  coupling to MS 74
  detection 73, 74, 76
  DNA analysis 69, 78, 83
  electrokinetic injection 72
  electropherogram 71
  elution order 75
  hydrodynamic injection 72, 73
  injection 72
  instrumentation 70, 76, 78
  limit of detection 74
  micellar electrokinetic chromagraphy 77
  modes 70
  optimisation of separation 76
  power supply 71
  protein analysis 69, 78, 82
  qualitative analysis 74
  quantitative analysis 74

sample stacking 73
separation principles 70
thermostatic control 72
capillary gel electrophoresis 70, 82
  cross-linked gels 83
  detection 83
  instrumentation 83
  linear gels 83
capillary isoelectric focussing 70, 76
  electroosmotic mobilisation 76
  electrophoretic mobilisation 77
  hydrodynamic mobilisation 77
  instrumentation 76
capture antibody 120
carboxy methyl 38
carboxypeptidase 172
carrier ampholyte 65
carrier electrolyte 72
cathodic mobilisation 77
cationic surfactants 61, 79
catholyte 66
CCD camera
  see charged coupled device camera
cDNA
  see complementary DNA
CE
  see capillary electrophoresis
cell lysis 144
centrifugation
  density 144
charged coupled device camera 140
chemical fragmentation methods 183
chemical ionisation 86
chemical mediator 128, 129, 130
chemical surface modification 53
chemiluminescent reactions 124, 136
cholesterol 121
chromatogram 29, 31
chromatographic theory 31
  capacity factor 32
  efficiency 32
  peak width 31
  plate height 32
  plate number 32
  resolution 33
  retention time 31
  retention volume 44
  selectivity factor 32
  van Deemter equation 32
  zero retention time 31
chromatography 22, 24, 25, 29
  affinity chromatography 34, 40
  amino acids 35, 37, 45
  analytical 37
  chromatogram 29, 31

detection methods 31, 36, 37
gas chromatography 30
general elution problem 33
gradient elution 33
high performance liquid chromatography
  36
high pressure liquid chromatography 36
instrumentation 36
ion exchange chromatography 34, 37
ion pairing reagents 35
isocratic elution 33
liquid chromatography 30
micellar electrokinetic chromatography
  77
micro 37
mobile phase 29, 30, 31, 34, 35, 39,
  45, 77
nano 37
normal phase 31, 34
peptides 35, 45
preparative 37
principle 29
proteins 37, 40, 45, 171, 174, 177
reversed phase 31, 34
size exclusion chromatography 34,
  42, 177
stationary phase 29, 30, 35, 38, 40, 43
chromosomal DNA 143, 144, 145
CIEF
  see capillary isoelectric focussing
citraconic acid 180
Clark, L.C. 125
cleavage
  disulfide bonds 175
  DNA 158
  immunoglobulins 112
  peptide bond 171, 174, 179, 180
  restriction enzymes 69, 143, 156, 157
Cleland's reagent 176
CM
  see carboxy methyl
CMC
  see critical micelle concentration
collagen 9
collision induced dissociation 104
complementary 19, 131
complementary DNA 134, 155, 166
convective diffusion 50
Coomassie brilliant blue 62
cortisone 113
coupling
  capillary electrophoresis - mass
    spectrometry 74
  liquid chromatography - mass
    spectrometry 37

Crick, Francis 18
critical micelle concentration 78
CsCl density gradient centrifugation 144
C-terminal analysis 171
  enzymatic methods 171
  hydrozinolysis 171
C-terminus 10, 170
cyanogen bromide 183
Cysteine 6, 13
Cystin 13
cytoplasm 20
Cytosine 15, 18

dansyl chloride 171, 179
ddNTP
  see 2′,3′-dideoxnucleotide triphosphate
DEAE
  see diethyl aminoethyl
degradation reactions 170
de-convoluted spectrum 103
denaturation
  DNA 19, 146
  proteins 14, 63, 64
denaturing agent 61, 63, 64, 65
densitometry 63, 82
deoxynucleotide triphosphate 137, 146,
  152, 163
deoxyribonucleic acid
  see DNA
detection
  amperometric 128
  diode array detector 37, 73
  electrochemical 128
  enzymatic reactions 122
  fluorescence 37, 74
  isotopic counting 119
  laser induced fluorescence 74, 164
  luminescence 119
  mass spectrometry 37, 74
  nephelometry 119
  optical 128
  refractive index 74
  turbidimetry 119
  UV absorption 31, 36, 73, 83
detergents 61, 65, 79, 145
  aggregation number 79
  anionic 61, 79
  as denaturation agents 61, 63, 64
  cationic 61, 79
  EOF control 53
  for bioassays 115
  for cell lysis 145
  for GE 61
  for IEF 65

for MEKC 78, 79
for PCR 153
non-ionic 79
SDS 61, 64, 78, 79, 115, 145
zwitter-ionic 61, 79
dextrose 83
dideoxynucleotide triphosphate 164
diethyl aminoethyl 38
diffuse layer 51
digestion 69
dimethyl sulfate 160
dimethyl sulfoxide 153
diode array detector 37, 73
dipeptide 9, 11
disulfide bond 9, 13, 111, 175
disulfide bond cleavage 175
disulfide bridge
  see disulfide bond
dithiothreitol 176
DNA 14, 143
  3D-structure 18
  base pairs 18, 19
  chromosomal 144, 145
  components 15
  denaturation 19
  density 144, 145
  double helix 18
  plasmid 144, 145
  primary structure 19
  secondary structure 19
  sequencing 156
  supercoiling 19, 145
  tertiary structure 19
  unfolding 19
DNA amplification
  see polymerase chain reaction
DNA arrays 25, 131, 156
  analysis 134
  applications 134, 136
  detection 132, 133
  fabrication 132
  fingerprint 132
  macroarray 133
  microarray 133
  principle 131
  sequencing 134
DNA binding arrays
  see DNA arrays
DNA chip 131
DNA extraction 143
  ClCs density gradient centrifugation 144
  total cellular DNA isolation 145
DNA microarray 131
DNA polymerase 122, 137, 146, 151, 163

DNA sequencing 22, 23, 25, 131, 134, 143, 156
  chain terminator method 162
  chemical cleavage method 158
  Dideoxy Method 162
  DNA arrays 131, 134
  Maxam-Gilbert method 158
  pyrosequencing 136
  Sanger method 162
dNTP
  see deoxynucleotide triphosphate
Dole, Malcolm 98
dose-response curve 116, 118
double helix 18
double stranded DNA 18, 19, 153
dsDNA
  see double stranded DNA
DTT
  see dithiothreitol
dynamic coating 53

Eddy diffusion 32
Edman degradation 170, 171, 175, 183
Edman's reagent 171, 179
efficiency 32, 148
electric double layer 50, 51
electric sector fields 86
electroendoosmosis 52, 59
electrokinetic chromatogram 78
electrokinetic injection 72
electrolyte 48
electron-impact ionisation 86
electron microscopy 24, 26
electroosmotic flow 48, 50, 52, 69, 75, 82, 83
  control 53
  flow profile 52
electroosmotic mobilisation 76
electroosmotic mobility 52
electropherogram 71, 73
electrophoresis 47
  2D-GE 67
  capillary electrophoresis 69
  capillary gel electrophoresis 82
  capillary isoelectric focussing 76
  gel electrophoresis 56
  isoelectric focussing 64, 76
  ladder 62, 64
  micellar electrokinetic chromatography 77
  principle 48
  SDS-PAGE 63
electrophoretic mobilisation 77
electrophoretic mobility 48, 49

electrophoretic theory 48
  apparent mobility 54, 56
  electroosmotic flow 48, 50
  electrophoretic mobility 48, 49
  Joule Heating 50
  migration time 55
  migration velocity 55
  peak dispersion 55
  plate number 55
  resolution 56
electrospray ionisation
  principle 98
  source 99
electrospray ionisation mass spectrometry 26, 37, 74, 86, 97
  applications 101
  comparison to MALDI-TOF-MS 106
  coupling to capillary electrophoresis 74
  coupling to liquid chromatography 37
  interface 99
  ionisation principle 98
  micro electrospray 100
  molecular weight determination 101
  nano electrospray 100
  negative ion mode 99
  peptide sequencing 101, 106
  pneumatically assisted electrospray 100
  positive ion mode 99
  quadrupole analyser 100
  structural analysis 104
ELISA
  see enzyme-linked immunosorbent assay
endopeptidase 179, 180
enzyme immunoassay
  see enzyme-linked immunosorbent assay
enzyme label 121, 122
enzyme linked immunosorbent assay 121
  enzymes 122
  HIV detection 124
enzymes 8, 42, 121, 122, 125, 126, 137, 151
  acetylcholine esterase 122
  alcohol oxidase 127
  alkaline phosphatase 122, 158
  aminopeptidase 171
  apyrase 137, 138, 139
  as bioreceptors 126
  ATP sulfurylase 137, 138
  carboxypeptidase 172
  classification 122, 123
  DNA polymerase 122, 137, 146, 151, 163
  endopeptidases 179, 180
  exopeptidases 171, 172, 179, 180
  function 8, 122
  glucose oxidase 127, 129, 130
  horseradish peroxidase 122, 123

enzymes (*cont.*)
  lactate oxidase 127
  luciferase 137, 138
  papain 112, 113
  polynucleotide kinase 158
  proteinase K 145
  restriction endonucleases 156
  restriction enzymes 69, 143, 156, 157
  reverse transcriptase 155
  ribonuclease 12, 145
  signal amplification 122
  trypsin 25, 96, 180, 181
  turnover 123
  urease 122, 127
EOF
  see electroosmotic flow
epitope 110, 113, 114
  continuous 114
  discontinuous 114
equilibrium constant 115
ESI
  see electrospray ionisation
ESI-MS
  see electrospray ionisation mass
    spectrometry
ethanol sensor 127
ethidium bromide
  CsCl density gradient 145
  real-time PCR 154
ethyl silane 35
exclusion limit 44
exonuclease activity 151, 154
exopeptidase 171, 172, 179, 180

F(ab)$_2$-fragment 113
Fab-fragment 112, 113
Faraday cup 86
fast atom bombardment 86
Fc-fragment 112, 113
Fenn, John 98
ferritin 8, 113
ferrocene 130
ferrocenium 130
fidelity 149
fingerprint
  2D-GE 67
  DNA arrays 132
  MALDI-TOF-MS 96
  peptide 96
flow profile 52
fluorescence detection
  DNA arrays 132
  immunoassays 119
  reversed phase liquid chromatography 37

Fluorescence Resonance Energy Transfer
    (FRET) 154
fluorophore 119
full width at half maximum 92
fused silica capillary 72

gas chromatography 30
GE
    see gel electrophoresis
gel electrophoresis 25, 26, 56, 82, 156, 157,
    160, 164
  2D-GE 25, 26, 67
  bands 57
  detection 62, 68, 83
  DNA fragments 156, 160, 164
  electrophoresis chamber 57
  gel media 59
  instrumentation 57, 65, 83
  isoelectric focussing 57, 64, 76
  native gel electrophoresis 56
  pore size gradient 61, 63, 69
  power supply 57
  protein fragments 177
  sample preparation 61
  SDS-PAGE 56, 63, 160, 177
  slab gel 57
  thermostat 57
gel filtration chromatography
    see size exclusion chromatography
gel permeation chromatography
    see size exclusion chromatography
gels 47, 48, 56, 57, 59, 63, 83
  agarose 48, 59
  chemical gels 83
  cross-linked 83
  degree of cross linking 59
  ion exchange chromatography 38
  linear 83
  non-restrictive 59
  physical 83
  polyacrylamide 48, 59, 60
  pore size gradient 61, 63
  restrictive 59
  total gel concentration 59
GeneChip 133, 136
general elution problem 33
genetic information 14, 143
genome 23, 131
genomics 23
geometric amplification 149
Gilbert, Walter 156
glucagon 9
gluconic acid 129
glucose 127, 129

glucose oxidase 127, 129, 130
glycerine 152
glycerol 57, 61
glycoprotein 20, 42
glycosilation 20
gradient elution 33
gradual desorption 39
gravity flow injection 73
groove binder
   see minor groove binder
group-specific ligands 42
Guanine 15, 18

haemoglobin 8
hapten 113
hCG
   see human chorionic gonadotropin
height equivalent of a theoretical plate
   see plate number
Hendersen-Hasselbalch equation 6
heterogenic 13
high performance capillary electrophoresis
   see capillary electrophoresis
high performance liquid chromatography
   see liquid chromatography
Hillenkamp, Franz 87
hinge region 111
HIV
   see human immunodeficiency virus
HIV test 124
home pregnancy test 120
homogenic 13
homopolymeric region 139
hormones 8, 42
horseradish peroxidase 122, 123
HPCE
   see high performance capillary
     electrophoresis
HPLC
   see high performance liquid
     chromatography
HRP
   see horseradish peroxidase
human chorionic gonadotropin 120, 121
human immunodeficiency virus 124
hybridisation 132, 134
   primer 137, 147, 154
hydrolysis
   proteins 178, 179
hydrazine 158, 160
hydrazinolysis 174
hydrodynamic injection 73
hydrodynamic mobilisation 77
hydrogenase 122, 123

IgG
   see immunoglobulin
immobilines 67
immobilised pH gradients 67
immune system 110, 111
immuno complex
   see antibody-antigen complex
immunoassays
   see bioassays
immunochemical complex
   see antibody-antigen complex
immunoglobulin 111, 112, 113
   classification 111
   F(ab)$_2$-fragment 113
   Fab-fragment 112, 113
   Fc-fragment 112, 113
   structure 111, 112, 113
immunosensor 127
insulin 9, 13
intercalating dye
   see intercalator
intercalator
   density gradient centrifugation 145
   real-time PCR 154
iodoacetic acid 176
ion exchange chromatography 25, 34,
     37, 179
   anion exchangers 38, 39
   buffers 39
   cation exchangers 38, 39
   gradual desorption 39
   stationary phase 38
ion pairing reagents 35
ionisation 86, 87, 98
IPG
   see immobilised pH gradients
IR spectroscopy 22, 24
isocratic elution 33
isoelectric focussing 57, 64, 76
   additives 65
   anolyte 66
   carrier ampholytes 65
   catholyte 66
   gels 65
   immobilins 67
   immobilised pH-gradient 67
   instrumentation 65, 76
   IPG gels 67
   mobilisation 76, 77
   pH drift 67
   pH gradient formation 65
   principle 64, 76
   resolution 65

isoelectric point 6, 39, 64, 77
  amino acids 7
  proteins 7
isolation
  DNA 24, 143, 144, 145
  proteins 25, 40
  RNA 24, 143, 145

Jorgenson, James 69
Joule heating 50, 56, 57, 69, 72

Karas, Michael 87
key-lock principle 8, 109
Krynn, Lukas 69

labels 110, 115, 117, 119, 122
lactate oxidase 127
lactose permease
  ESI-spectrum 103
ladder 62, 64
laser induced fluorescence detection 74, 164
laser light scattering detection 119
lasers 119
latex particles 119
limited proteolysis 181
lipoprotein 20
liquid chromatography 22, 24, 25, 26, 30
  affinity chromatography 40
  analytical 37
  applications in bioanalysis 34
  coupling to MS 26, 37, 102
  detectors 31, 36, 37, 45
  ion exchange chromatography 37
  micro 37
  nano 37
  normal phase 31, 34
  preparative 37
  reversed phase 31, 34
  size exclusion chromatography 42
longitudinal diffusion 32
luciferase 137, 138
luciferin 137, 138
Lyons, C. L. 125

macroarray 133
magnesium chloride 152
MALDI-TOF-MS
  see matrix assisted laser desorption
    ionisation time of flight mass
    spectrometry
mass spectrometry 22, 25, 37, 74, 85, 177,
  187
  chemical ionisation 86

detector 85, 86, 92
electron impact ionisation 86
electrospray ionisation 86, 98
fast atom bombardment 86
hard ionisation 86
ion sources 85, 86
ionisation techniques 86
mass analyser 85, 86, 90, 100
matrix assisted laser desorption ionisation
  86, 87
principle 85
quadrupole analyser 86, 87, 100
resolution 92, 93, 101
soft ionisation 86
tandem mass spectrometry 101, 104
time of flight analyser 90
mass to charge ratio 85, 86, 90, 101
mass transfer 32
matrix 87, 88, 93
matrix assisted laser desorption ionisation
  87, 88, 89
matrix assisted laser desorption ionisation
  time of flight mass spectrometry 25,
  86, 87
  applications 94
  comparison to ESI-MS 95, 106
  dried droplet method 93, 94
  fast evaporation method 94
  flight time 90, 91
  function of matrix 93
  ionisation principle 87
  lasers 89
  linear analyser 90, 91
  mass to charge ratio 90
  matrix materials 87, 88
  molecular weight determination 94, 96
  post soure decay 91
  reflectron analyser 91
  resolution 92, 93
  sample preparation 93
matrix materials 87, 88
Maxam-Gilbert method 25, 143, 156, 158
mediator
  see chemical mediator
MEKC
  see micellar electrokinetic
    chromatography
mercaptoethanol
  see 2-mercaptoethanol
Merrifield resin 185
messenger RNA 14, 20, 143, 155
metabolites 30
methanol 80
methylation 20

micellar electrokinetic chromatography 70, 77
  applications 77, 78
  capacity factor 81
  critical micelle concentration 78
  detergents 78, 79
  elution ratio 80
  instrumentation 78
  migration window 80
  modifiers 82
  optimisation of separation 81
  principles 78
  resolution 81
  selectivity factor 81
  theory 81
micelles 77, 78
  aggregation number 79
  co-micelles 82
  critical micelle concentration 78
Michaelis-Menten equation 123
microarray 133
microparticles 119, 140
microspheres
  see microparticles
migration time 55
migration velocity 55
minor groove binder 154
mobile phase 29, 30, 31, 34, 35, 39, 45, 77
molecular ion 89
molecular recognition 8, 109
  affinity chromatography 40
  bioassays 110
  biosensors 125
  DNA arrays 131
  pyrosequencing 136
molecular sieving 48, 56, 59, 64
molecular weight determination
  capillary gel electrophoresis 83
  ESI-MS 101
  MALDI-TOF-MS 94, 96
  nucleic acids 25, 45, 63, 83, 101, 160, 164
  pore gradient gel electrophoresis 63
  proteins 25, 45, 63, 64, 83, 94, 96, 101, 177
  SDS-PAGE 57, 63
  size exclusion chromatography 45
monoclonal 112
monospecific ligands 42
mRNA
  see messenger RNA
Mullis, Kari 146
multichannel plate 86
multi-determinant 113

nanoparticles 119
nebulising gas 99
nephelometry 119
neurotensin
  ESI-spectrum 102
NMR spectroscopy 22, 24, 26
non-specific bonding 115
normal phase liquid chromatography 31, 34
N-terminal analysis 171
N-terminus 10
nucleic acid amplification 146
nucleic acid sequencing 25, 131, 134, 136, 143, 156
  Maxam-Gilbert (chemical cleavage) method 158
  Sanger (chain terminator or dideoxy) method 162
  use of restriction enzymes 156
nucleic acids 14, 15, 143, 146, 156
  3'-end 18
  5'-end 18
  analysis 23, 143
  backbone 18
  base pairs 18, 19
  components 15, 16, 17
  extraction and isolation 24, 143, 144, 145
  nucleobases 15, 16
  nucleoside 15, 16
  nucleotide 16, 17, 151
  PCR 146
  phosphate 15, 16
  purine 15, 16
  pyrimidine 15, 16
  Sanger method 25, 143, 156, 162
  short forms 18
  side groups 18
  structure 15, 18, 20
  sugars 15, 16
  transcription 20, 21
  translation 20, 21
nucleoside 15, 16
nucleotide 16, 17, 151
number of theoretical plates
  see plate number

octadecyl silane 35
octyl silane 35
ODS
  see octadecyl silane
oligopeptide 9
OPA
  see ortho-phthalaldehyde
ortho-phthalaldehyde 179
oxido-reductase 122, 123

PA
  see polyacrylamide
PAGE
  see polyacrylamide gel electrophoresis
papain 112, 113
paratope 110, 112, 113, 114
PCR
  see polymerase chain reaction
PEG
  see polyethylene glycol
Pelletier elements 148
pepsin 113, 181
peptide bond 7, 9, 10
  acid-catalysed hydrolysis 178
  base-catalysed hydrolysis 179
  chemical fragmentation 183
  enzymatic cleavage 171, 172, 179, 180
peptide fingerprint 96, 97
peptides 1, 7
  structural analysis 104
performic acid 176
Pfu-polymerase 151
pH drift 67
pH gradient 64
  immobilised 67
  linear 67
  non-linear 67
phenyl isothiocyanate 171
phosphoric acid 15
phosphorylation 20
physiological pH 4
phytochelatin
  tandem MS-spectrum 105
pI
  see isoelectric point
piperidine 160
PITC
  see phenyl isothiocyanate
plasmid DNA 143, 144, 145
plate height 32
plate number 32, 55, 81
plug flow profile 52
polyacryl amide 48, 59, 60, 64, 83
polyacrylamide gel electrophoresis
  see sodium dodecyl sulfate
    polyacrylamide gel electrophoresis
polyclonal 112
polyethylene glycol 53, 83, 115, 153
polymer networks 83
polymerase
  see DNA polymerase
polymerase chain reaction 24, 25, 143, 146
  additives 152
  buffers 152, 153
  cycle 148

  denaturation 146
  DNA polymerase 51
  efficiency 148
  exponential amplification 149
  fidelity 149
  magnesium chloride 152
  plateau phase 149
  primer 152
  primer annealing 146
  primer extension 147
  principle 146
  rate of amplification 149
  reagents 151
  real-time PCR 24, 153
  reverse transcription 155
  specificity 149
  steps 146, 147
  temperature control 148
  thermal cycling 148
  thermus aquaticus 151
polymerisation activity 152
polynucleotide 17
polynucleotide kinase 158
polypeptide 9
pore size gradient electrophoresis 63
post source decay 91
post-translational modification 20, 64, 95
pregnancy test 120
pressure injection 73
primary antibody 117, 118
primary structure
  DNA 19
  proteins 9
primer 137, 146, 152, 163
protein sequencing 26, 169
  amino acid composition 169, 178
  cleavage by aminopeptidases 171
  cleavage by carboxypeptidases 172
  C-terminal analysis 172
  disulfide bond cleavage 175
  disulfide bond position 186
  Edman degradation 170, 171, 175, 183
  Edman's reagent 171
  end group analysis 170
  enzymatic methods 171, 172, 179, 180
  gas phase 185
  hydrazinolysis 174
  mass spectrometry 104, 187
  N-terminal analysis 171
  ordering of peptide fragments 186
  solid phase 185
  specific peptide bond cleavage 179
  strategy 170
proteinase K method 145
proteinase K 145

proteins 1, 7, 25, 169
  α-helix 11
  acid catalysed hydrolysis 178
  amino acid composition 169, 170, 178
  analysis 25, 63, 64, 68, 83, 94, 101, 109,
    169
  antibodies 8, 111, 127
  antigens 8, 113
  backbone 10
  β-folding 12
  β-plated sheet 12
  chemical fragmentation 183
  chromatographic separation 34, 38, 39,
    45
  C-terminal analysis 172
  C-terminus 10
  denaturation 14, 61, 63, 65, 175
  disulfide bond cleavage 175
  disulfide bridges 9, 13
  Edman-degradation 170, 171, 175, 183
  electrophoretic separation 63, 64, 68, 76,
    83
  end-group analysis 170
  enzymatic fragmentation 178, 180
  enzymes 8, 122, 126
  fibrous 9
  folding 9
  function 8
  gas-phase sequencing 185
  globular 8, 9
  heterogenic 13
  homogenic 13
  isoelectric point 6, 7, 39, 64, 77
  isolation 25
  molecular weight determination 25, 45,
    64, 67, 83, 94, 101
  N-terminal analysis 171
  N-terminus 10
  pI values 7
  post-translational modification 20, 64, 95
  primary structure 9
  purification 25
  quantitative analysis 25
  quaternary structure 13
  re-naturation 14
  secondary structure 11
  sequencing 169, 183
  side-chains 10
  structure 9
  synthesis 20
  tertiary structure 12
  tryptic digestion 25, 96, 97, 180, 182
proteolytic enzyme 145
proteomics 25
Purine 15

Pwo-polymerase 151
Pyrimidine 15
pyrogram 138, 139
pyrophosphate 137
pyrosequencing 25, 136, 156
  applications 140
  instrumentation 140
  principle 137
  reagents 137
  sample preparation 140
  single nucleotide polymorphism 140

quadrupolar magnetic field 87
quadrupole analyser 100
quasi-molecular ion 89
quaternary structure
  proteins 9, 13
quencher 154

radioisotope 119
real time PCR 24, 153
  dsDNA binding dye assay 153
  probe based assay 154, 155
recognition site
  see molecular recognition
reflector TOF 91
reporter 154
residues 10
resolution
  10 % valley 92
  50 % valley 92
  chromatography 33
  electrophoresis 56
  full width at half maximum 92
  MEKC 81
  isoelectric focussing 65
  mass spectrometry 92
response curve
  see dose-response curve
restriction endonculease 156
restriction enzyme 69, 143, 156, 157
restriction map 157
retention time 29, 31, 44
reverse transcriptase 155
reverse transcription 134, 143, 155, 166
reversed phase liquid chromatography 31,
    34, 102, 179
  buffers 35
  instrumentation 36
  ion pairing reagents 35
  mobile phase 35
  stationary phase 35
r-hirudin
  MALDI-TOF-spectrum 95

ribonuclease 8, 12, 145
ribonucleic acid
   see RNA
ribosome 20
RNA 8, 14, 15, 20, 24, 69, 131, 143, 145,
   166
   density 144
   extraction 24, 143, 145
   messenger RNA 14, 20, 143, 155
   sequencing 166
   structure 15, 20
   transfer RNA 15, 20
   two dimensional gel electrophoresis 69
RNA-directed DNA polymerase
   see reverse transcription
Ronaghi, Mostafa 136
RT
   see reverse transcription
run buffer 48

salt mobilisation 77
sample stacking 73
sandwich assay 118, 120
sandwich complex 117, 120
Sanger method 25, 143, 156, 162
Sanger, Frederick 156, 170
scintiallation counter 87
SDS
   see sodiumdodecyl sulphate
SDS-PAGE
   see sodiumdodecyl sulphate
      polyacrylamide gel electrophoresis
secondary antibody 117, 118, 124
secondary electron multiplier 87
secondary structure
   DNA 19
   proteins 11
selectivity 40
selectivity factor 32, 56
   in MEKC 81
separation
   amino acids 25, 30, 31, 34, 35, 37, 45, 64,
      78, 179
   charged compounds 37, 70, 75, 77
   DNA 25, 31, 37, 45, 69, 78, 82, 83, 160,
      164
   neutral species 24, 70, 75
   proteins 26, 31, 34, 35, 37, 40, 45, 57, 64,
      68, 83, 96, 171, 174, 177, 179, 187
   RNA 31, 37, 45, 69, 78, 82, 83
   zwitter-ionic compounds 64, 70, 76
separation methods
   affinity chromatography 40
   capillary electrophoresis 69
   capillary gel electrophoresis 82

gas chromatography 30
gel electrophoresis 56
ion exchange liquid chromatography 37
isoelectric focussing 64, 76
micellar electrokinetic chromatography
   77
reversed phase liquid chromatography 34
SDS-PAGE 63
size exclusion chromatography 42
TOF-mass spectrometry 96
two dimensional gel electrophoresis 67
sequenator 183
sequencing
   see DNA sequencing
   see protein sequencing
   see RNA sequencing
side chain 10
side groups 18
sieving effect
   see molecular sieving
signal amplification 122
signal transduction 125
silica particles 35
silver staining 62
sinapinic acid 87
single nucleotide polymorphism 110, 131,
   136, 140
single stranded DNA 19, 132, 137, 140,
   146ff, 157
size exclusion chromatography 34, 42ff, 144
   applications 45
   buffers 45
   calibration curve 44
   exclusion limit 44
   isolation 144
   molecular weight determination 45
   stationary phase 43
   total permeation 44
   total volume 44
slab gel
   see gels
SNP
   see single nucleotide polymorphism
sodiumdodecyl sulphate 61, 64, 78, 79, 115,
   145
sodiumdodecyl sulphate polyacrylamide gel
   electrophoresis 25, 63ff, 67, 177
somatostatin 9
specificity 40, 110, 126, 149
ssDNA
   see single stranded DNA
stacking
   see sample stacking

stationary phase 29, 30
  affinity chromatography 40
  ion-exchange liquid chromatography 38
  reversed phase liquid chromatography 35
  size exclusion chromatography 43
Stern layer 51
Sudan III 80
sulfide bridge
  see disulfide bond
surfactants
  see detergents

Tanaka, Koichi 87
tandem mass spectrometry 104
  daughter ion analysis 104
  scanning mode 104
  static mode 104
Taq polymerase 151
Terabe, Shigeru 77
tertiary structure
  DNA 19
  proteins 12
TFA
  see trifluoroacetic acid
theophylline 113
thermal cycling 148
Thymine 15, 18, 20
time of flight analyser 86, 87, 90ff
  linear 90, 91
  reflectron 91
Tiselius, Arne 47
TOF analyser
  see time of flight analyser
total permeation 44
tracer antibody 120
transcription 20, 21
transducer 125, 127ff
transfer RNA 15, 20
transferrin 8
translation 20, 21

trifluoroacetic acid 36
tripeptide 9
Triton X-100 61, 79, 82, 144
tRNA
  see transfer RNA
trypsin 25, 96, 180, 181
tryptic digestion 25, 96, 97, 180, 182
Tsweet, Mikhail Semenovich 29
Tth-polymerase 151
turbidity 119
turnover 123
Tween-20 153

Uracil 15, 20
urea 122, 127
urease 122, 127
urine glucose 121
UV spectroscopy 22
  nucleic acids 24
  proteins 25
UV-detection
  capillary electrophoresis 73, 83
  CsCl density gradient centrifugation 145
  liquid chromatography 31, 36
  size exclusion chromatography 45

vacuum injection 73
valinomycin 11
van Deemter equation 32

Watson, James 18

x-ray crystallography 22, 24, 26

z-cell 73, 74
zero retention time 31
zeta potential 52
zwitter-ionic 6, 61, 79

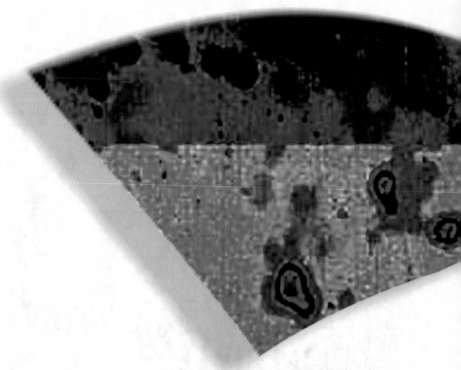

bioanalytical chemistry

I nterdisciplinary knowledge is becoming more and more
important to the modern...
textbook covers bioanaly... mainly the
analysis of proteins and DNA) an... explains everything for the

nonbiolog...
bioassays,...
in conver...
describe...
instrumer...
to chemis...
knowledg...

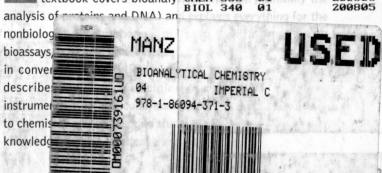

Imperial Colle...

www.icpress.co.uk

Out
of the
Barn

The Instrumentation, Systems, and Automation Society (ISA)

by Dick Morley